ACTION SCIENCE
UNIT THREE

Cover image: Eruption of Mount Rinjani, Indonesia, with simultaneous lightning strike. Image © Oliver Spalt, CC

Back Cover: U.S. Marine Corps / Public Domain

ACTION SCIENCE
UNIT THREE: THE FLOW OF ENERGY

Peter F. Swan
New Haven, CT

Copyright © 2005, 2021 by Peter F. Swan

Orders: www.amazon.com

All rights reserved. No part of this book may be reproduced or transmitted in any form or by any means, electronic or mechanical, including photocopying, recording, or by any information storage and retrieval system, without written permission from the author, except for the inclusion of brief quotations in a review.

ISBN: 9798520634904

To Dianne

Thank you for supporting my goals

Available energy is the main object at stake in the struggle for existence and the evolution of the world.
— Ludwig Eduard Boltzmann

The laws of light and of heat translate each other; so do the laws of sound and color; electricity and magnetism are varied forms of this selfsame energy.
— Ralph Waldo Emerson

Acknowledgements:

I wish to thank the following dedicated educators for the time and effort they contributed to read and comment on this text: Amy Bell, Collegiate School NYC; Kathleen Daly, Milford Public School Science Coordinator, Milford, CT; Violette Barasch, Easton, CT; John Cunningham, Foote School, New Haven, CT; Roderick Mobley, Choate Rosemary Hall, Wallingford, CT; Sandra Bethray, King & Low-Heywood Thomas School, Stamford, CT; Eileen Glassmire, Cheshire Academy, Cheshire CT; Cathie Bischoff and Sophie Homan, Rye Country Day School, Rye, NY; Carolyn Kraemer, The Williams School, New London, CT. In addition, those teachers and administrators of Hopkins School who provided encouragement, support, and feedback for this project deserve my special gratitude: Barbara, Ben, Bruce, Carla, Jennifer, Katie, Sally, and Thach.

Contents

1 Heat Energy — 1

   1.1 What is Energy? — 1
   1.2 Types of Energy — 4
   1.3 Energy Transformations — 6
   Practice Problems — 7
   1.4 Working Safely with Heat — 8
   Experiment: Crank the Heat — 11
   1.5 Units of Heat — 16
   Extending Ideas — 19
   Activity: The Color of Motion — 20
   Experiment: Food Energy — 22
   1.6 Food Energy — 24
   Extending Ideas — 25
   1.7 Temperature Scales — 26
   1.8 Absolutely Freezing! — 28
   Practice Problems — 29
   Extending Ideas — 30

2 Equilibrium and Phases — 31

- 2.1 Exchanging Heat — 31
- 2.2 Heat and Temperature — 34
- Practice Problems — 35
- Extending Ideas — 36
- Activity: Compromise! — 37
- Practice Problems — 39
- 2.3 An Equilibrium Equation — 40
- Practice Problems — 42
- Extending Ideas — 43
- Experiment: Specific Heat — 44
- Table: Specific Heat — 47
- Practice Problems — 48
- Extending Ideas — 49
- 2.4 Phase Changes — 50
- Activity: Going Through a Phase — 52
- 2.5 Latent Heat — 54
- Extending Ideas — 58
- Table: Latent Heat — 58

3 Heat Energy and Efficiency 59

3.1 Thermodynamics	59
Experiment: Chill Out, Man!	62
3.2 Thermal Expansion	65
Practice Problems	69
Extending Ideas	70
3.3 Heat Transfer	71
3.4 Energy and the Earth	76
Extending Ideas	79
3.5 Heating and Cooling Systems	81
Extending Ideas	86
3.6 Engines and Power Plants	87
Activity: Efficiency	88
3.7 Consumption and Conservation	90
Practice Problems	92

4 Heat and Light — 93

- 4.1 Energy in Motion — 93
- 4.2 Describing Waves — 95
- Activity: Mini Hendrix — 98
- Practice Problems — 99
- Extending Ideas — 101
- 4.3 Charge Vibrations — 102
- Activity: Electromagnetic Slide — 104
- 4.4 The Electro-Magnetic Link — 105
- 4.5 Sound and Light — 108
- Activity: The Light We Cannot See — 111
- 4.6 The Speed of Dark — 113
- 4.7 Doing the Wave — 114
- Extending Ideas — 121
- 4.8 Analog vs. Digital — 122

Illustration / Photo Credits — 125
Index — 127

1
Heat Energy
1.1 What is Energy?

Before you get dressed in the morning, you might want to know the **temperature** outside. If you are cold while inside, you could adjust the thermostat; this will raise the temperature, but this will use more **energy**. Of course, in the summer, you may wish to turn on the air conditioner if you can't bear the **heat**. This will lower the temperature – but will also use more energy.

You are certainly familiar with the three words in bold text, and probably use these ideas almost every day. Like many terms used in science, these have specific meanings that are not always understood in everyday conversation. What exactly do these words mean? What is the link between temperature, heat, and energy?

We will start with perhaps the least understood of these terms, energy. This idea has a central place in society today, but the meaning was only defined clearly in the middle of the nineteenth century. The concept of energy was not accepted the same by all scientists until the industrial revolution was well underway.

> TEMPERATURE, HEAT and ENERGY are familiar terms. Precise definitions help us understand how they are related.

> ENERGY is the capacity to perform work. A helpful example of work is the ability to lift a weight.
>
> POTENTIAL ENERGY is energy that is stored in some way, but can be transformed into work.

That origin gives a clue to our definition: **energy** is the capacity to do work. A technical definition of *work* is not necessary at the present time, so we will simply note that one obvious sign of the capacity to do work is that a weight can be lifted. Often, energy is stored – not actually doing any work at the moment, but able to do so when needed. This is called **potential energy**, and it is not always easy to observe. A compressed spring and a full gas tank each hold potential energy. A flashlight battery stores energy, and so does your lunch. When stored energy is used, it does not disappear, but simply changes from one kind of energy to another.

Can the above examples of potential energy perform work by lifting a weight? The spring can extend and push a small weight upwards. The gasoline can enable a car to drive up a hill. It is not as easy to see how the flashlight battery can lift a weight, but suppose we remove the battery from the flashlight and attach it to a small electric motor instead. Now the stored energy can be used to wind a string and lift a weight! Although the motor is used to actually perform the work, the energy stored in the battery makes it possible. Similarly, your lunch certainly can't lift a weight by itself – but you can't work for long without the energy that food provides!

The test for the presence of energy is whether it *can possibly be* turned into work. There are many forms of energy, and actually lifting a weight is not required to prove that it is present. The spring can simply stay compressed; it still holds potential energy. The gasoline can remain unburned; in fact, it can remain in the ground as crude oil - never pumped to the surface, never refined and delivered to a gas station, never put into a car. The energy stored is no less real. The battery can stay in the flashlight, changing its stored energy into light instead of work. As we will learn, light is another form of energy. Many people eat lunch every day although they have no intention of lifting any weights. The food contains energy that will be used by the body or stored as fat until needed.

Energy can be used to perform work. The ability to lift a weight is a useful definition of work.

1.2 Types of Energy

The term "energy" is important both in science and in your everyday life. In fact, understanding and using energy is one of the most important applications of science. You need energy every day to do everything you do. Your body obtains energy from food and uses most of this food energy simply to keep your heart, lungs, brain, and other internal organs functioning properly. Of course, energy in other forms is used to heat and cool your home, to help you travel from place to place, and to run the many electrical appliances that have become essential to the modern lifestyle.

There are many different types of energy and many different sources of energy. One sign of a chemical reaction is that energy can be released, often observed by a temperature change. A microwave oven uses electrical energy to warm up your dinner. Energy has a central place in our lives, yet we do not often look at where it comes from, how it is used, and what happens to it after we use it!

Understanding that last statement is puzzling, although it might not seem to be. You may think that once you have used energy, it is gone – that it has ceased to be. Careful measurements show that this is not true – energy can change from one type to another, but it never disappears from the universe.

The rest of this book will explore how energy can change, and how you can observe and measure those changes. First, we need to define the different kinds of energy, so that we can have some idea of what we are looking for. There are several main categories of energy, which can be further broken into many specific types.

Mechanical energy describes the action of masses and forces, such as a spring or a moving mass. **Radiant energy** can travel long distances very quickly; light is an example. **Electrical energy** relates to the forces and motion of charged particles. **Chemical energy** involves changes in the form of matter; when gasoline burns or food is digested, chemical energy is released. **Thermal energy** describes the motion of atoms or molecules within a substance. **Nuclear energy** comes from changes in the atomic nucleus of atoms; a tremendous source of stored energy that we are still learning to control. Within these categories, there are several different specific types. You encounter energy in dozens of different forms every day.

The food that you eat goes through chemical reactions within your body. Energy is released when organic compounds break down and form into new compounds. *Metabolism* is the term for the complex chemical and physical processes that produce the energy needed to sustain life.

1.3 Energy Transformations

Because energy is not always active, it is not always obvious. We are usually most aware of energy when it is changing from one type to another. For example, when atoms form molecules in chemical reactions, energy is released or absorbed in the position of electrons in their chemical bonds. We only observe this chemical energy when the bonds break and re-form into other molecules. The difference in energy of the two compounds is sometimes seen by a release of heat or light.

Scientists today have clear definitions of the many forms of energy. One of the most important scientific laws stems from our understanding of the many transformations that energy can undergo:

The Law of Conservation of Energy states that energy is never created or destroyed, but often changes from one type to another.

When energy is converted, part of the energy becomes less useful as it changes from one type to another. Some of the energy always converts to a less concentrated form that is more difficult to convert into work. The energy spreads out, as the current of a mighty river disperses when the river widens into a bay and the water flows gently into the ocean. So it is with energy; each change releases some energy as **heat**, which tends to disperse as time passes, spreading out and less useful for doing work.

Thermal energy that spreads out or moves from one material to another is called HEAT.

Energy is continually changing form; nuclear energy in the sun and all other stars constantly transforms into radiant energy. The portion of this energy that reaches earth changes into chemical energy or heat. Stored chemical energy on earth is converted through many processes, both natural and man-made. The energy becomes less and less concentrated, and some turns into heat with every conversion. Scientists think that eventually all of the energy in the universe will be distributed equally throughout space. With all of the energy spread evenly, there can be no more energy conversions; the eventual end of energy changes is called the "heat death" of the universe. Don't worry too much about that; it won't happen anytime soon!

Practice Problems

1) Name three types of energy that you use or observe every day. You may name a general type as described on page 5, or a specific kind. Explain the energy transformation that you can observe while using this energy.

2) Name three ways that energy can be stored. Describe one way that each of these examples of potential energy can be transformed into another type of energy.

3) Describe three energy transformations that do not produce useful work, as it is defined in this chapter.

Energy can easily transform into many different types, and it is most easily observed during such a change. This winch on board a fishing boat is used to haul in the nets – work that would otherwise depend on human muscle power! A generator on board the boat changes gasoline (chemical energy) into heat (radiant energy) to run a generator (mechanical energy) that produces electricity (radiant energy). The electricity runs a motor (mechanical energy) that lifts the load into the boat (work).

1.4 Working Safely with Heat

The previous section discussed how heat tends to spread out or *dissipate* as time passes. That will cause some challenges in the lab, as you try to measure the temperature of a substance that is cooling off. That means you will need to plan carefully and work efficiently, and it also means that often your temperature readings will be a bit different than you might expect.

A more immediate concern is the danger of *concentrated* heat energy. In short: you can be burned or scalded by contact with a hot substance. You must be aware of that fact in all labs and activities in this section. Although you are always required to behave in a responsible, safe manner in science lab, there are some special precautions that you should take during heat energy labs.

You may use a hot plate, microwave oven, stovetop or candle as a heat source. The danger of an open flame is obvious; do not reach over a flame, and be sure to secure or remove all loose clothing and tie back all long hair before starting the lab. Hot plates or stovetops can also present a danger. They may or may not glow red (electric) or have a visible flame (gas). These heat sources may look the same whether on or off, and will remain hot for some time even after they are turned off. Even an unplugged hot plate may have been recently used and therefore may still be very hot. ALWAYS assume that the hot plate or stovetop is hot, and be very cautious of touching any part other than the controls.

You must also work efficiently to get accurate measurements while a substance is heating or cooling. Remember, heat energy tends to spread out as time passes. This is why you should always be well-organized and prepared before starting any lab.

You will use a thermometer or temperature probe to measure temperature; be careful! It is a delicate instrument; if placed on the lab table, it may roll off and break. If left upright in a container, it may fall over. With care, you will easily avoid this kind of mishap.

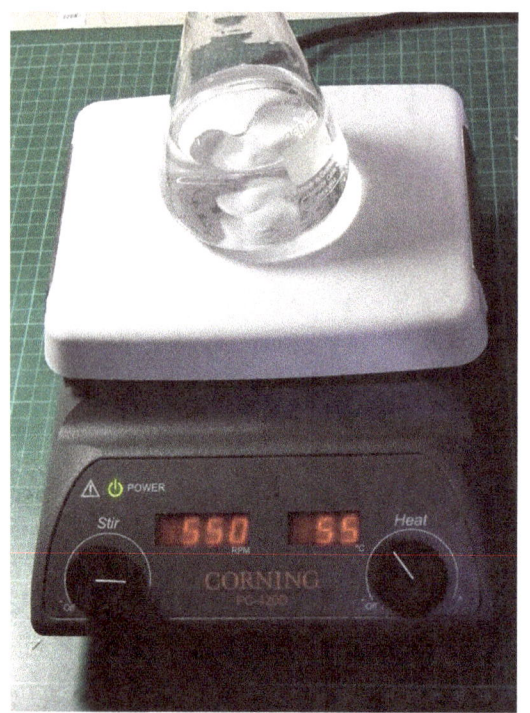

If you use gas burners or hot plates in the lab, be sure that you understand how to use them safely. Never reach over a flame, and do not go near a flame with loose clothing or hair. Always assume that a burner or plate is hot, even if it is not turned on. The metal remains hot for some time after the gas or electricity has been turned off.

You may use glass beakers and flasks; they will break if dropped, tipped over, or mishandled. Be cautious; never leave a thermometer standing in a beaker or cup because the weight of the thermometer may tip the whole thing over, and there will be hot water and broken glass in your work area.

Keep in mind the main idea of lab safety: the science lab is a safe place to work, but not a safe place to play. You should not be fearful, but do be careful, aware of the dangers, and thoughtful in your actions as you carry out your tasks.

Experiment: Crank the Heat

Purpose: This experiment will examine the relationship between heat, energy, and temperature. You will add heat energy to different amounts of water using different heating devices, then measure the change in temperature for each sample.

Materials: room temperature water   thermometer
graduated cylinder   beaker
heat-resistant gloves   heating devices

Procedure: In general, you will measure the amount of water, the starting temperature, and the temperature as you heat samples of water by various methods. Detailed instructions are provided in each part below. Every time that you measure temperature, stir the sample with the thermometer for 10 to 20 seconds before reading the temperature. This will help the water temperature become consistent throughout the sample, and will also allow time for the thermometer to read the correct temperature.

Always handle beakers carefully, using tongs or gloves if necessary. When you are finished with all heating and temperature readings, empty the warm water into the sink or container supplied for that purpose (not into the room-temperature water pitcher).

There are several parts to this lab. You may do them in any order, to make the equipment available to all students. Answer the questions for each part right after you do the experiment. **Read each part carefully, and construct a data table before you start.**

Data: Part 1: You will first heat 100 mL of water, then repeat with 200 mL. Your table should look like this:

Time:	0 sec.	10 sec	20 sec	30 sec	40 sec	50 sec	60 sec
100 mL Temp:							
200 mL Temp:							

Measure 100 mL of water using the graduated cylinder, then pour it into the beaker. Measure and record the water temperature. Use the microwave to heat the water for 10 seconds. Measure and record the water temperature.

Put the beaker back in the microwave, and heat it for another 10 seconds. Again, measure and record the water temperature. Repeat 4 more times so that you have the starting temperature and 6 more temperatures, each after heating for 10 more seconds.

Pour out the hot water, and allow the beaker to cool. Then measure 200 mL of water, and repeat the process.

Analysis: Part 1:
1) Make a scatter graph showing the temperature on the vertical axis. You will use the same graph to record the data for both amounts of water, so include a temperature range suitable for both. The time is on the horizontal axis, from 0 seconds to 60 seconds. Plot the data set for 100 mL of water, and join the data points with a single straight line that seems to fit the data best. Do the same with the data set for 200 mL of water.

2) The microwave oven changes electrical energy into radiant energy, which changes into heat energy when it hits the water. Is heat energy being added to the water during the entire time that you tested? Explain your answer, using the graph.

3) Compare the graphs for the two data sets. Be specific about how the graphs change during the 60 seconds. Does the temperature change at the same rate for the different amounts of water, or does one change temperature faster than the other?

4) Do you think that the same amount of electrical energy transforms into heat energy during each of the 60-second periods? Describe evidence from your graph or your data to support your answer.

Data: Part 2: Measure 100 mL of water using the graduated cylinder, then pour it into the beaker. Measure and record the water temperature. Turn on the heating device (hot plate, stovetop, or burner) to the setting designated by your teacher, and give it some time to heat up. This may take a few minutes, which you may use to prepare a data table. Your table should look like this:

Time:	0 min	1 min	2 min	3 min	4 min	5 min	6 min
100 mL Temp:							
200 mL Temp:							

Place the beaker on the heating device, and measure the temperature every minute. You will need to work carefully with a partner, with one partner stirring, measuring and saying the temperature, and the other partner keeping track of the time and recording the data in the table. Continue to measure and record the temperature every minute for 5 minutes of heating.

Analysis: Part 2:

5) Make a scatter graph showing the temperature on the vertical axis. You will use the same graph to record the data for both amounts of water, so include a temperature range suitable for both. The time is on the horizontal axis, from 0 minutes to 6 minutes.

   Plot the data set for 100 mL of water, and join the data points with a single straight line that seems to fit the data best. Do the same with the data set for 200 mL of water.

6) An electric hot plate or stove-top changes electrical energy into heat energy. A gas burner or gas stovetop changes chemical energy into heat energy. Was heat energy being added to the water during the entire time that you tested? Explain your answer, using the graph.

7) Compare the graphs for the two data sets. Be specific about how the graphs change during the 6 minutes. Does the temperature change at the same rate for the different amounts of water, or does one change temperature faster than the other?

8) Do you think that the same amount of energy transforms into heat energy during each of the 6 minute periods? Describe evidence from your graph or your data to support your answer.

After you have completed both parts 1 and 2, and answered all questions 1 through 8:

9) Which device added heat energy faster during your tests? Describe evidence from your graph or your data to support your answer. Why is it important to use the same sample size and heating time for this comparison?

10) If you add the same amount of heat energy to a cup of water or a pot of water, which will have a larger change in temperature? Explain your answer.

Now you will use the data and the graphs to make some predictions. First, select one of the heating devices that you used:

11) What would be the temperature of a 100 mL sample heated for twice as long as your experiment? State which heating device, and show the work that you use to make this prediction.

12) What would be the temperature of a 400 mL sample heated for the same amount of time as your experiment? State which heating device, and show the work that you use to make this prediction.

13) In your own words, describe the relationship between temperature, heat, and energy as demonstrated in this experiment.

1.5 Units of Heat

It is always difficult to define something that you cannot see. The definition of energy was open to debate until the middle of the nineteenth century. Heat, a form of energy, was also a mysterious term, difficult to pin down. At one time, scientists thought that heat is a fluid that can move between objects. They called it "caloric," which comes from the Latin word for heat, and is the origin of our modern term for the energy in food. Experiments later forced them to abandon this model.

Heat energy is added to the air in a balloon.

When scientists created a practical meaning of energy, they could define heat as a form of energy, because heat has the capacity to do work. A weight can be lifted by adding heat to a balloon filled with air; this is what happens in a hot-air balloon. This is not a very efficient method to use heat, but it does satisfy the definition of energy. Our modern definition of heat describes what happens when we add heat energy to matter.

All substances are made of atoms, and these tiny particles can move and interact with one another. Gas particles are quite free to move in all directions. The molecules of a liquid are less mobile; they can move around within the limits of a constant volume. Solids do not change their shape, so the particles cannot move around, but only vibrate in place.

In each case, adding energy causes the particles to move more quickly. On a scale much too small to see, heat energy is increasing the motion of atoms and molecules. When you touch something that is hot, its molecules have a lot of motion. All molecules are constantly vibrating, and in liquids and gases they are moving around as well. The molecules transfer some of this motion to your hand, which you sense as heat. Too much heat is painful, which is your body's way to warn you before your skin suffers a burn from exposure to heat.

> Molecules of all kinds of matter are constantly in motion. In solids, this motion is vibration only; in liquids and gases, the motion is also from place to place.

All kinds of energy can be measured in units called **joules**, named after nineteenth-century British scientist James Joule (pronounced like the word "jewel"). He experimented with energy conversions and clearly stated the law of conservation of energy. Because we cannot see heat, Joule learned by studying the *effects* of heat. He performed important experiments and analysis on the relationship between heat and work. Joule showed by experiments that heat is a form of energy, and he demonstrated that work and energy are the same.

Many other units of energy are commonly used, such as ergs, Btu, and calories. All are applied to specific uses. Ergs are a metric unit used only for extremely small amounts of energy. Btu (British thermal units) are often used in heating or air-conditioning systems of buildings. Calories with a capital "C" (also called kilocalories) are commonly used to measure the energy content of food.

Any of these can be converted to the SI energy unit of joules. The original SI unit for heat was the calorie, equal to the amount of heat needed to raise one gram of water by one degree Celsius. After James Joule showed that heat, energy, and work are different forms of the same thing, the conversion was found to be about 4.2 joules in one calorie.

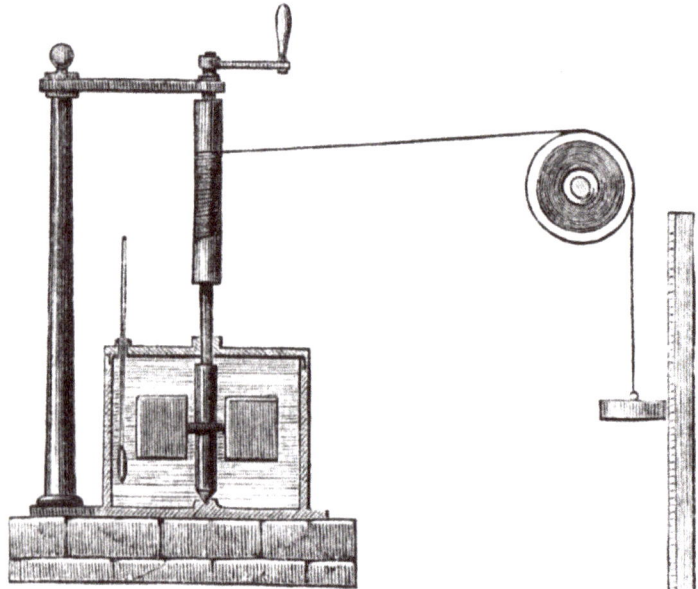

James Joule's diagram for his experiment. The weight at the right falls and pulls a string. The string unwinds, spinning the paddles in the water contained in an insulated container at left. A thermometer in the water measures temperature change. A crank winds the string for another trial.

Extending Ideas

1) A diagram of James Joule's experiment comparing work to heat energy is on the previous page. Sketch this diagram, and label each part that is in bold letters in the description below.

   The **weight** falls and pulls a **string**. The string unwinds, spinning the **paddles** in the **water** contained in an **insulated container**. A **thermometer** in the water measures temperature change. A **crank** winds the string back up.

2) Use your understanding of work, energy, and temperature to fill in the blanks. Possible answers are listed in the Word Bank. You may use answers more than once, and you will not need all of them.

   a) If twice as much weight is used, the work done by the falling weight would be _____ as much.
   b) If the length of the string were doubled, the paddles would spin _____ as much.
   c) If there were two times the amount of water in the container, the heat energy added would be _____, and the temperature change would be _____ as much.
   d) James Joule's experiment shows that work and energy are _____.

Word Bank: the same   different   twice   half

3) In the diagram of Joule's experiment, turning the crank lifts the weight, which means that work is done. Do you think this increases the water temperature? If so, where does that energy come from?

Activity: The Color of Motion

Almost two centuries ago, a Scottish botanist named Robert Brown observed the motion of tiny lifeless particles under a microscope. He described the motion, which no one could adequately explain until Albert Einstein did so nearly a century later. We now understand this "Brownian Motion" is caused by collisions of these tiny particles with molecules – much, much smaller, but moving so fast that the effect can be seen under high magnification. We can never see these molecules moving; but like Mr. Brown, perhaps we can observe an *effect* of this motion. We won't even need a microscope!

Materials: three beakers   food coloring
                 microwave oven or hot plate   thermometer
                 electronic balance or scale   water

Fill the three beakers about half way with water. Allow the water to settle for a few minutes, as you carefully measure the mass of each one. They should be similar, but they do not need to be exactly the same. Place all three beakers on the same piece of paper, and label them Cool, Warm, and Hot (even though right now they are all the same temperature). Below the labels, write the mass of each one. You may use these papers to record all of the data for this lab.

Place the "Warm" and "Hot" beakers in the microwave, and turn it on for 30 seconds. Remove the "Warm" beaker and measure and record its mass. Meanwhile, turn the microwave back on for another 30 seconds with the "Hot" beaker inside. Then remove the "Hot" beaker, measure and record its mass, and place all three beakers in their places on the paper. Starting with the "Cool" beaker, measure and record all three temperatures.

Carefully add one drop of food coloring to each beaker, one right after the other. You may use the same color for all three, or different colors if you prefer. Do not stir or jostle the beakers in any way. Observe the spread of the food coloring through each beaker.

1) Does heat have mass? Describe evidence from the experiment to support your answer.
2) In which container are the molecules of water moving fastest? Describe evidence from the experiment to support your answer.

Careful experiments never show an increase in mass when heat is added to a substance. Since all matter has mass and volume, heat is not matter.

Experiment: Food Energy

The energy content of food is measured in the unit of Calories. Scientists originally measured this amount for different foods in a way similar to that used in this experiment, although they used more sophisticated equipment. The numbers you get for this experiment will not be very accurate, but the important part is that you understand the method and the concept. The other important part is that you do not burn yourself!

Materials: popcorn   2 paperclips   dinner plate
matches   water   digital scale
coffee cup with handle   digital thermometer
100 mL graduated cylinder

Data: Mass of popcorn, starting and ending temperature of water

Procedure: Measure 100 mL of water, and pour it into the coffee mug. Measure and record the mass of two pieces of popcorn. Bend the paper clips to hold the popcorn as shown below, and poke a piece of popcorn onto each one. Set them on the dinner plate.

Measure the temperature of the water in the mug. Hold the mug over the popcorn, and carefully light the popcorn on fire. Warm the water until the popcorn completely burns (you may need to relight it if it goes out). Then measure the temperature of the water again.

Analysis: Answer the following questions.

1) What happened to the temperature of the water? Where did this energy come from?

2) Subtract the starting temperature of the eater from the final temperature to find the change in temperature (you will learn that this is ΔT (called "delta T", or change in temperature).

3) It takes 1 calorie to heat 1 gram of water by one degree Celsius. Multiply 100 grams by the ΔT. This is the number of calories you measured in two pieces of popcorn.

4) Divide the number of calories that you calculated by the mass of the popcorn that you measured before burning. This is the calories per gram of popcorn.

5) A lot of heat energy escaped and did not go into the water. You probably felt some of this as you held the mug by the handle. Some heat also warmed up the mug itself, instead of the water. Do you think this means that the actual "calories per gram" of popcorn is *higher* or *lower* than what you calculated?

1.6 Food Energy

Everything that you eat or drink contains chemical energy. A chemical reaction that produces heat is called **exothermic**. This happens whenever something burns in the presence of oxygen, but it also happens at a much slower, more controlled rate as your body digests food. This process is called **metabolism**. Although different methods are used now, this is how scientists once measured the amount of calories in food. They used a device called a "bomb calorimeter" to be sure that no heat was lost to the surroundings, and they used methods to be sure the food burned completely. The bomb calorimeter did not explode (unless something went very wrong!), but it did contain high pressure, which led to its cool name.

> EXOTHERMIC reactions produce heat energy.
>
> METABOLISM is the process that breaks down the food that you eat, extracting the energy your body needs.

When your body digests food, similar chemical reactions occur, but at a much more controlled rate as you *metabolize* everything that you eat or drink. The Calories in food are just a measure of the thermal energy given off by the chemical reaction of combining organic material with oxygen. It does not matter whether this reaction occurs in your body, in a bomb calorimeter, or in a science lab; the amount of energy is the same! Incidentally, the Calorie with a capitol "C" indicates 1,000 calories with a lower-case c; this is also called a kilocalorie.

Extending Ideas:
Breakfast, Lunch, and Dinner

Based on your understanding of the potential energy in food, write a short story that illustrates your use of food energy during a typical day. Try to be realistic about the amount of energy that you take in during meals or snacks, and how much you use during various activities. Be creative, but not too lengthy. One to two pages is plenty.

Food energy is usually measured in Calories. The capital "C" means that a Calorie is 1,000 calories (a kilocalorie). This is about 4,200 joules, or 4.2 kilojoules. A typical student consumes about 6,000 to 10,000 kilojoules per day. You may use either unit, but be consistent throughout your story. Simply estimate your total food energy intake in a day, then divide it reasonably among your meals and snacks.

Remember that a day is 24 hours long. Your body needs energy even while sleeping – to keep your heart beating, your blood flowing, and all of your organs working properly (including the brain – yes, you do burn more calories studying or taking a test than you do while watching TV!). Depending on your size (bigger people need more energy) this could be from 40 to 100 Calories (168 to 420 kilojoules) per hour. Of course, you consume much more energy when you are active. For example, a hard soccer game might require 800 Calories.

You are at an age when you are growing! If you are ten pounds bigger this year than you were at the same time last year, that works out to about an extra 100 Calories per day (420 kJ) to help your body grow! Other than that, your input energy should be the same as your output energy; remember conservation of energy!

1.7 Temperature Scales

James Joule measured temperature changes in his experiments using a thermometer marked off in degrees Fahrenheit. This scale is still widely used in the United States. It is named after a German scientist who invented the mercury thermometer in 1714. Gabriel Fahrenheit mixed salt, water, and ice to attain the coldest temperature that he could in his science lab; he marked this as zero degrees. He established 32 scale marks higher as the freezing point of water; 32 degrees Fahrenheit.

There are several different stories about why the scale is set up like this. He tinkered with his scale divisions, possibly trying to make the 100 degree mark correspond to human body temperature, so that scientists everywhere could reproduce the scale. Fahrenheit wanted anyone in the world to be able to make a thermometer and mark it with the same scale. You are certainly familiar with the Fahrenheit scale, where water boils at 212 °F, and 70 °F is a comfortable room temperature.

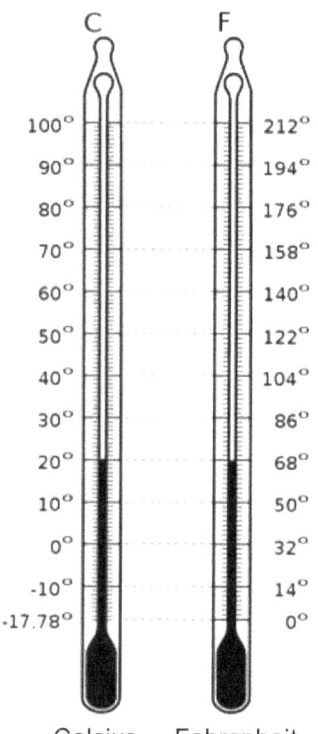

Gabriel Fahrenheit invented the mercury thermometer in 1714, and he worked out a scale that could be reproduced in any science lab. The Celsius scale is now used in most countries around the world.

The SI thermometer is based on water. The freezing point of water is designated as zero, and there are 100 divisions until the boiling point is reached. The name "degrees" is retained from the old scale, so water boils at 100 degrees Celsius, named for Anders Celsius, the Swedish scientist who first suggested the idea. You may hear people refer to its original name, the *centigrade* scale, but Celsius is now the official term. Fortunately, both centigrade and Celsius start with the same letter, so the unit "°C" is understood either way.

You saw that adding energy to water increases the **temperature**. We cannot measure heat directly, but temperature is a measure of the motion of atoms and molecules. James Joule's work helped explain temperature changes in terms of the motion of these particles. When **heat energy** is added, molecular motion increases. The energy of any motion is called **kinetic energy**, after the Greek word for movement. The change in motion of atoms and molecules depends on how much heat is added, and the number, mass, and freedom of movement of these particles.

KINETIC ENERGY is energy of motion. TEMPERATURE is a way to measure the kinetic energy of molecules. HEAT is the name of energy that changes this motion.

1.8 Absolutely Freezing

The temperature scale used in many science experiments is the Kelvin scale. The zero point of this scale is where theories show that all motion of molecules stops. At this point, there is no molecular kinetic energy; so there is no heat. This point is called "absolute zero", and it has never been reached in the lab, although scientists have come very close. Helium turns to a liquid at 4.2 K, and temperatures of less than 1 K have been reached in experiments. Zero on this scale, usually called "absolute zero," is about 273 degrees below zero Celsius!

Hydrogen must be chilled to 20 kelvins to change phase to a liquid. Along with liquid oxygen (chilled to 90 K), it is used as rocket fuel. The "smoke" that you see in this rocket engine test is actually steam, as the hydrogen and oxygen combine to form water and give off a tremendous amount of heat energy!

So is zero kelvin the coldest possible temperature? The problem with that statement is that as absolute zero is approached, we can no longer remove that last bit of heat energy. Heat energy can flow, but that requires a temperature difference. Since heat only flows from warmer to cooler, there is no way to remove that last bit of energy; there is no "colder" place for it to flow *to*. So you would be correct to say that absolute zero is the coldest *impossible* temperature!

The size of one kelvin is the same as one degree Celsius, so a change of 10 kelvins is the same as a change of 10 °C (notice that the units are not called "degrees kelvin" but just "kelvins"). Because most of our experiments observe a *change* in temperature rather than the actual number above or below zero, we can use either scale.

Practice Problems

Use the side-by-side thermometers shown in section 1.7 to answer questions 1 and 2. Use information in section 1.8 to answer 3 and 4.

1) The text says that 70 °F is a comfortable room temperature. What is that approximate temperature in degrees Celsius?

2) Average internal temperature of the human body is about 99 °F. What is this in degrees Celsius?

3) What is the freezing point of water in kelvins?

4) What is a comfortable room temperature in kelvins?

Extending Ideas

1) An old theory suggested that heat is a type of fluid that can flow from one substance into another. This was called the "caloric" theory. Describe results from one or more of the experiments you have done in this chapter that provide evidence against this theory.

2) Sometimes an experiment predicts that you will not measure a change. This is called a "null hypothesis." Describe a measurement in one experiment from this chapter that predicts a null hypothesis.

3) Is it possible that a null hypothesis is confirmed by an experiment, but the measuring instruments are not sensitive enough to measure a change that actually occurred?

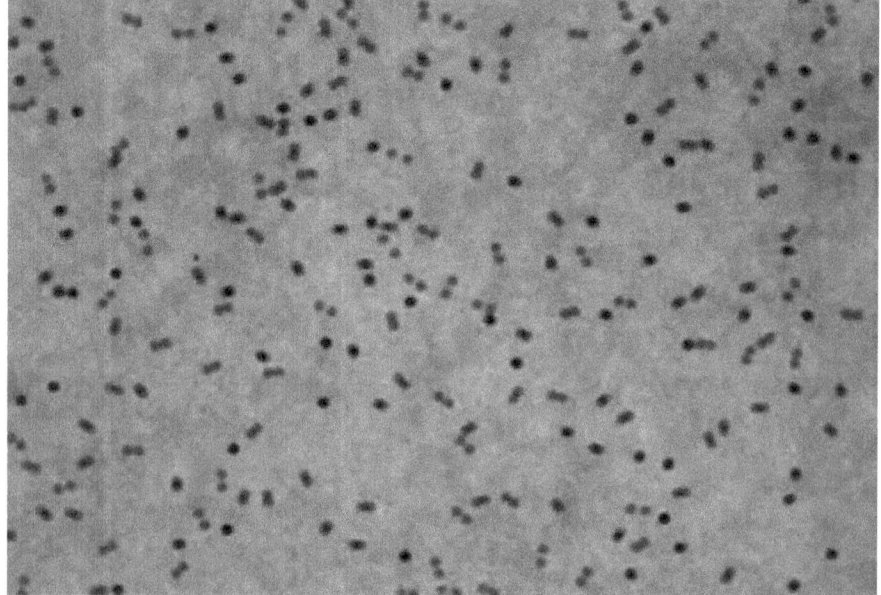

Adding heat increases the motion of atoms and molecules. This helps heat energy spread around and move from one substance to another. You are not seeing actual atoms and molecules here, but a simulation to help picture this sub-microscopic activity.

2
Equilibrium and Phases
2.1 Exchanging Heat

You have used a microwave oven or hot plate to heat food or water, changing electrical energy into thermal energy. You may also have experience using a gas stove, oven, or grill. The most common gasses used in these devices are methane (often called "natural gas") and propane (often called "LP gas"). This is completely different from the "gas" that is put in a car, which is actually short for "gasoline." All of these devices – grills, ovens, and car engines - transform chemical energy into thermal energy.

But what happens to the energy then? If you heat water, does it stay hot? Does it turn back into chemical or electrical energy? Of course not – a hot cup of tea will cool off as it sits. Where does that energy go? The conservation of energy law tells us that it does not disappear, but it does *dissipate* – that means spread out. So the energy from the hot tea goes into the air in the room.

A hot mug of tea loses heat energy to the air. The temperature of the room increases just a little bit.

You found in your experiments that you must work efficiently when measuring temperature. Some of the heat energy escapes from the container as time passes. Heat energy tends to spread out whenever possible. The energy becomes less structured instead of remaining concentrated in one location. We could use an insulated cup to reduce the spread of heat energy to the air in the room. This only slows down the process; after a while, even water in an insulated cup cools to room temperature. As the water cools, the air in the room warms a little bit to compensate. The water and the air will soon reach the same temperature.

The law of conservation of energy helps make sense of just what happens in this situation. Substances in contact can freely exchange heat energy when those moving molecules collide. As time passes, they will tend to reach an equal temperature. Since energy cannot be created or destroyed, the heat lost by one substance must be gained by another. They are said to reach **thermal equilibrium**.

THERMAL EQUILIBRIUM: constant temperature. No further heat exchange is measurable.

The term Equilibrium describes any situation where there is no observable change. This condition applies to many situations in chemistry and physics.

Thermal, from the Greek word for heat, is also the root word for *thermometer*. The substance that has a higher temperature loses energy to the cooler substance. When two or more samples reach the same temperature, the exchange of heat is no longer observed.

The condition of equilibrium simply indicates that a constant or steady state is reached, and it occurs in many situations other than heat exchange. A chair sitting on the floor is in equilibrium, and a spaceship gliding through space is also in equilibrium. If you spill a glass of water on the floor, the water will spread into a puddle. The effect of gravity on fluid prevents the water from remaining in a pile, but the water has internal forces that tend to hold molecules together. The puddle does not spread forever, but will reach a point of equilibrium where these opposing influences balance.

Your body is constantly losing heat to the surrounding air, so you might think that you will reach room temperature after a while. However, you have an internal source of energy – the food that you eat is *metabolized* (chemically transformed) by your body into energy that keeps your body at the proper temperature of about 37°C. Your body is in equilibrium because it has an equal amount of energy flowing into it and out of it. As you can see, equilibrium does not mean nothing is happening; but that effects of opposing influences are balanced. A constant state is reached and no further change is evident.

The thermostat in your house turns on the furnace automatically. Heat from the furnace offsets the various ways that heat escapes your home. The temperature inside your home stays the same. Equilibrium does not mean that nothing is happening, but that loss and gain are balanced.

2.2 Heat and Temperature

We need to use indirect methods to determine both heat and temperature. We cannot see atoms and molecules to measure their motion, so we find temperature changes instead. A simple equation relates these properties in quantities that we can measure:

$$\Delta H = mc\Delta T$$

The triangle is a Greek letter "delta", and indicates a change in quantity. The **m** stands for mass of the substance being heated. The letter **c** represents a property of the substance called **specific heat**. The specific heat is the amount of energy needed to change the temperature of one gram of a substance by one degree Celsius. The heat capacity of water is 4.2 joules per gram per degree, written as $c = 4.2$ J/g·°C. The **T** indicates temperature, so the formula can be stated in plain language as follows:

> The SPECIFIC HEAT is the amount of energy needed to change the temperature of one gram of a substance by one degree Celsius. For water, that amount is about 4.2 joules.

Change in heat energy is equal to the mass of a substance times its specific heat times the change in temperature.

As an example, suppose you want to raise the temperature of a cup of 200 grams of water from 20 °C (room temperature) to 45 °C to make some tea. The heat energy you must add is:

200 g x 4.2 J/g·°C x 25 °C = 21,000 joules of energy.

Try that calculation to be sure that you get the same answer. As you can see, the mass and temperature units cancel out, leaving the answer in joules. You might wonder where the 25 °C comes from. Remember, it is not the *temperature*, but the *change in temperature* that is important. The change of a quantity is always calculated by subtracting the initial value from the final value. For example, if you start the day with eight dollars and have two dollars left in the evening, the change in money is: $2 - $8 = -$6. The change or "delta" is negative six dollars. Using subscripts to indicate *initial* and *final*, the change in temperature is calculated:

$$\Delta T = T_f - T_i$$

The "delta" or change in a quantity is always found by subtracting the final from the initial amount. Subscripts in the equation help to identify which is which.

Practice Problems

Find the change in temperature for each of the following:
1) Water cools from 45 °C to 12 °C.
2) Water warms from 20 °C to 80 °C.
3) A block of ice warms from -24 °C to -9 °C.
4) A block of steel cools from 20 °C to -8 °C

Find the change in heat energy for each of the following:
5) 40 grams of water cools from 45 °C to 12 °C.
6) 80 grams of water cools from 45 °C to 12 °C.
7) 80 grams of water warms from 20 °C to 80 °C.
8) 40 grams of water warms from 20 °C to 80 °C.

Extending Ideas

1) 40 grams of water is heated from 4°C to 20°C. How much heat energy has been added?

2) 120 grams of water starts at room temperature (20 °C). What is the final temperature of the water after adding 25,000 joules of heat energy?

3) Two samples of water start at to 20 °C and are heated in separate beakers. One beaker contains 40 grams of water, and it is heated to a final temperature of 70 °C. The other beaker contains 60 grams of water, and it is heated to a final temperature of 60 °C.
   a) Your book defines temperature as a measure of the motion of atoms and molecules. Since both samples contain molecules of water, decide which molecules (small or large sample) had the greater motion after heating; explain your reasoning.
   b) Which sample has had more heat energy added to it? Show all work clearly.
   c) If both samples were heated on the same hot plate at the same setting, which took more time to get to its final temperature? Explain your answer.

4) A small space heater is sometimes used in the home. Correctly use the terms <u>energy</u>, <u>temperature</u>, and <u>heat</u> to describe what this device does.

Activity: Compromise!

In this activity we will combine water of different temperatures; you can probably make an educated guess at the result, based on your experience adjusting the water every time you take a bath or shower. You will measure water in mL in this experiment. To convert to grams, remember that one mL of water has a mass of one gram.

Materials: thermometer insulated cup
warm and cold water graduated cylinder

Obtain your samples of warm and cold water from the sources as directed by your teacher. Get 50 mL of cold water. Pour it into the insulated cup and stir with the thermometer for about 30 seconds to measure and record the temperature.

Get 50 mL of warm water, and measure and record the temperature while it is still in the graduate; then pour it into the foam cup. Stir for about 30 seconds with the thermometer, then measure and record the temperature of the combined samples.

1) What happened to the temperature and heat energy of <u>each</u> sample when the warm and cold water were combined? (The warm water did *this* and the cold water did *that*)

2) Use the equation for change of heat energy, $\Delta H = mc\Delta T$, to calculate the heat gained by one sample and the heat lost by the other sample. Clearly indicate which calculation you are doing.

Repeat the experiment, using 25 mL of cold water and 75 mL of warm water. The temperatures of each may have changed since you last measured them, so make all measurements again.

3) Record the temperature of the cold water, the temperature of the warm water, and the temperature of the combined samples.

4) Use the equation for change of heat energy to calculate the heat gained by one sample and the heat lost by the other sample.

Repeat the experiment again, using 75 mL of cold water and 25 mL of warm water. Check all initial temperature measurements again.

5) Record the temperature of cold water, the temperature of warm water, and the temperature of the combined samples.

6) Use the equation for change of heat energy to calculate the heat gained by one sample and the heat lost by the other sample.

7) Look over your results and draw some conclusions from this experiment. Remember that a little heat may have been lost or gained from the air in the room.

Practice Problems

1) Based on your experience in the lab, make an *educated guess* of the final temperature when the following samples are mixed:
    a) 20 mL of water at 20 °C with 20 mL of water at 60 °C.
    b) 20 mL of water at 20 °C with 40 mL of water at 60 °C.
    c) 40 mL of water at 20 °C with 20 mL of water at 60 °C.
    d) 5 mL of water at 20 °C with 95 mL of water at 60 °C.
    e) 20 mL of water at 60 °C with 40 mL of water at 60 °C.
    f) 20 mL of water at 40 °C with 40 mL of water at 20 °C.

2) 20 mL of water at 10 °C is poured into a beaker with 30 mL of water at 45 °C. The water is stirred, and the final temperature is 29.5 °C.
    a) How much heat energy was lost by the hot water?
    b) How much heat energy was gained by the cold water?
    c) How much heat energy was lost to the surrounding air?
    d) Did this experiment violate the law of conservation of energy? Explain your answer.

Adjusting the hot and cold faucets mixes the right amounts of each for a comfortable equilibrium temperature. Fortunately, you don't need to do the math every time you take a bath!

2.3 An Equilibrium Equation

Two samples of water that start at different temperature will reach thermal equilibrium soon after being mixed. After sitting for a while, the mixed sample comes to equilibrium with the surrounding air. If you wish to find the equilibrium temperature of the original mixture, you must measure the sample before it cools to room temperature. An insulating cup will delay this process, but some heat will still disperse into the air. In the short term, this creates a small error in all of your heat experiments. In the long term, the universe tends toward overall equilibrium.

> ENTROPY is a measure of the disorder of a system. Among the many effects of entropy is the tendency for energy to disperse evenly.

In other words, energy tends to move from a concentrated, ordered condition to a dispersed, disorganized state. This tendency is known as **entropy**; although we will not study it in detail, you can use it as an excuse next time you get in trouble for having a messy room! The idea of entropy explains why the randomly moving gas particles in the air spread throughout a room, instead of collecting in one corner. Entropy also influences the heat energy of those particles; the heat becomes dispersed around the room even if there is a single heat source in one place. Hot particles near the heat source move quickly and carry that energy to all parts of the room, until the air in the room reaches equilibrium.

You can use the equation for heat energy and the idea of conservation of energy to predict the equilibrium temperature. For two mixed samples, this can be expressed in this mathematical equation. Remember that **ΔH** means change in heat energy:

ΔH of sample one + ΔH of sample two   = zero   or
$(mc\Delta T)_1$ + $(mc\Delta T)_2$ = 0

The change in heat energy for each sample is calculated just as you have done in previous exercises. One sample will lose energy, and one sample will gain energy. The total change in energy must be zero, as long as you use an insulated cup and measure the mixed sample quickly. The final temperature for both samples is the equilibrium temperature. If you know the mass and specific heat of each sample, you can solve the equation to find the equilibrium temperature.

$$m_1 c_1 (T_f - T_i)_1 + m_2 c_2 (T_f - T_i)_2 = 0$$

This equation may look complex, but if you are careful to write down the equation and fill in the known values carefully, it is not difficult to solve. With this equation, you can calculate the expected results of your previous activity. You should find that the calculations are a fairly close match to the actual experimental outcome. Despite some error due to the energy lost to the air in the room, this confirms the law of conservation of energy, as represented by the equilibrium equation.

Subscripts in the equilibrium equation are used differently than the subscripts in a chemical formula. These subscripts simply identify which sample is measured by each variable. That is, the mass of sample one is indicated by m_1, and the mass of sample two is indicated by m_2. Subscripts also indicate the *final* and *initial* temperature.

Practice Problems

1) A comfortable hot bath or shower is about 40 °C. If the hot water in your home is 50 °C and the cold water is 16 °C, how much cold water should be mixed with every liter of hot water to produce a comfortable bath temperature?

2) How much 50 °C water must be added to one liter of 20 °C water to reach an equilibrium temperature of 30 °C?

3) How much 20 °C water must be added to 160 mL of boiling water to reduce the temperature to 65 °C?

4) Suppose that you boil some water to make tea, and pour some water into a teacup. You immediately measure the temperature, and find that the water is only 88 °C, even though it was boiling at 100 °C in the teapot only seconds before. Explain what has happened to decrease the water temperature so quickly. Think of as many reasons as you can.

5) 200 mL of water at 40 °C is mixed with 200 mL of water at 60 °C.
   a) Guess the equilibrium temperature.
   b) Calculate the equilibrium temperature.

6) 300 mL of water at 40 °C is mixed with 100 mL of water at 60 °C.
   a) Guess the equilibrium temperature.
   b) Calculate the equilibrium temperature.

Extending Ideas

1) 10 mL of water at 8 °C is poured into an insulated container, and 20 mL of water at 40 °C is added.
   a) Find the equilibrium temperature: assume no heat energy is lost to the surroundings.
   b) Find the equilibrium temperature when 220 Joules of energy has been lost to the air in the room.

2) 100 mL of water at 40 °C is mixed with 300 mL of water at 60 °C.
   a) Guess the equilibrium temperature.
   b) Calculate the equilibrium temperature.

3) 40 mL of water at 30 °C are put in a container. 80 mL of 60 °C water is added and the water is stirred. The temperature is immediately measured and found to be 48 °C.
   a) How much heat energy was gained by the cool water?
   b) How much heat energy was lost by the hot water?
   c) What has happened to the "lost" energy?

4) Review your actual results from the "Compromise!" activity. Use the mass and initial temperature data for each set of warm and cold samples to predict the equilibrium temperature that the mixed samples should reach. Compare the calculated answer to the actual measured temperature of the experiment. Explain your results.

Experiment: Specific Heat

So far, water is the only substance that you have used in heat exchange equations and experiments. Water has a very high specific heat; this simply means that it takes a lot of heat energy to change its temperature. Because it is also very common and very chemically stable, it is useful in heat experiments. How can we study the effects of heat energy on other materials? The general equations for heat energy and equilibrium are still valid, because these equations are based on the law of conservation of energy. However, specific heat is a property. Remember that a property is a descriptive trait that helps to identify a substance. That means that you would expect that each substance has a different value of specific heat. The properties of water remain a useful basis for comparison of other substances. How does the specific heat of some common materials compare to that of water?

You will get a different kind of challenge in this experiment: you will need to create the method yourself. You will be provided with all of the materials that you need to find the heat capacity of a specific substance. You must first create a clear and effective procedure to carry out the experiment. Some ideas to keep in mind:
- If your cold substance starts at room temperature, it will not gain or lose any heat to the air.
- It is a good idea to measure your hot substance just before mixing, so that it does not have too much time to lose heat to the air.

- Mix and measure the substances in an insulated container. The substances must come to an equilibrium temperature, but must be measured before too much heat is lost to the surroundings.
- Before you start the experiment, think about the most efficient ways to measure the mass and temperature of the substances.
- Use the equilibrium equation to identify all measurements that you must make during the experiment. Your written procedure should describe exactly when and how to measure each item. Create a data table to record each measurement.

Write a lab report that includes the following headings. The italics must be replaced with the correct information for your experiment. **Your written procedure must be approved by your teacher before you actually start the experiment.**

Purpose: Design and carry out an experiment to find the specific heat of (*selected substance*)

Materials: water   thermometer
                 insulated cup   graduated cylinder
                 heating device   electronic balance

Procedure: *Describe your method to find each variable in the equilibrium equation. Add the selected substance to your materials list.*

Data: *Create a table to record all information.*

Analysis: Show the complete equilibrium equation, with all data written in (include units of all measurements). *Solve the equation to find the unknown heat capacity. Your teacher may require that you repeat the procedure, gather a new set of data, and verify the results.*

Post-lab questions:

1) What is your experimental result (the value that you calculated for specific heat)? If you repeated the experiment, your result is the average of your calculated values.

2) Ask your teacher for the value of specific heat of your substance. Compare this to your experimental results in a short paragraph.

3) When an experimental value is to be compared to a known or "accepted" value, the *percent difference* is often calculated. The percent difference is calculated this way:

percent difference = $\dfrac{\text{accepted value - experimental result}}{\text{accepted value}} \times 100\%$

Calculate the percent difference of your experimental result from the accepted value provided by your teacher.

Specific Heat of Some Common Materials

Material	Specific Heat
Aluminum	0.900 J/g•°C
Air	1.01 J/g•°C
Brass	0.370 J/g•°C
Copper	0.385 J/g•°C
Gasoline	2.09 J/g•°C
Glass	0.833 J/g•°C
Glycerin	2.43 J/g•°C
Gold	0.130 J/g•°C
Ice	2.00 J/g•°C
Iron	0.452 J/g•°C
Lead	0.130 J/g•°C
Marble	0.88 J/g•°C
Mercury	0.138 J/g•°C
Nickel	0.440 J/g•°C
Oak	2.38 J/g•°C
Olive oil	1.67 J/g•°C
Pine	2.80 J/g•°C
Sand	0.816 J/g•°C
Sea water	0.94 J/g•°C
Silver	0.234 J/g•°C
Titanium	0.582 J/g•°C
Water	4.19 J/g•°C
Zinc	0.276 J/g•°C

Practice Problems

(use the heat capacity table to help solve these problems)

1) A 90 gram piece of silver starts at room temperature (20°C). 800 joules of heat energy are added; what is the final temperature of the silver?

2) How much heat is needed to change 30.0 grams of ice at -10.0 °C into ice at -2.00 °C?

3) A piece of stainless steel with a mass of 3.88 grams absorbs 355 joules of heat energy when its temperature increases from 25.0 °C to 203.0 °C. What is the specific heat of stainless steel?

4) Suppose a 50 g block of unknown material is at room temperature of 20°C. The block is placed in an insulated container that holds 120 mL of water at 70 °C. After stirring, the equilibrium temperature is found to be 65 °C.
   a) How much heat energy was lost by the water?
   b) How much heat energy was gained by the block, if no heat escaped the container?
   c) Find the specific heat of the block.

Extending Ideas

(use the heat capacity table to help solve these problems)

1) Find the equilibrium temperature when 30 grams of copper at 60 °C is put into an insulated cup with 30 grams of 10 °C water.

2) Find the equilibrium temperature when 12 grams of aluminum at 40 °C is put into an insulated container with 50 grams of 22 °C water.

3) 250 grams of an unknown metal at 20 °C is placed in a well-insulated container with 40 mL of water at 5 °C. The equilibrium temperature is found to be 10.5 °C. What is the unknown material?

4) 50 mL of water at 80 °C is placed in a well-insulated container with a 40 gram block of metal that starts at 20 °C. The final temperature is found to be 72 °C. Assume that no heat is lost to the surroundings.
    a) What is the specific heat of the metal?
    b) If less water had been used, would the final temperature have been warmer or colder? Explain.

5) A 30 gram piece of aluminum at 200 °C is placed in an insulated container with 150 mL of room temperature (20 °C) water. Find the final equilibrium temperature if no heat energy is lost from the container.

2.4 Phase Changes

As you have discovered, specific heat is a property that varies from one substance to another. This characteristic can be tested to help identify an unknown sample, and this test does not change the substance. These facts mean that specific heat is a *physical property*. Like density or solubility, it can be *quantified*; that is, analyzed and assigned a numerical value.

You may live near a lake or pond, or you may pass one from time to time. Perhaps you have seen the pond on a cold day in late fall. The temperature may be less than 0 °C, but the pond has not turned to ice. Certainly you realize that water put into the freezer does not immediately turn to ice! Even though the temperature is below the freezing point of water, it takes time for this transformation to occur. The ice cubes may need a few hours; the pond may require a week of below-freezing weather.

Consider the reverse situation. Spring arrives, with warmer weather; yet the pond remains frozen for days. Or you take an ice cube from the freezer and drop it into a drink at room temperature. The cube slowly shrinks, yet remains visible for quite some time as it cools your drink. Again, in both cases the change is not immediate, even though the temperature is above the melting point.

If you add heat to water, eventually it will turn to water vapor. Depending on how quickly you add the heat, this process is either evaporation or boiling, and it may take minutes or days. If you leave a cold drink out on a hot day, the container gets wet. This is caused by water vapor in the air turning into liquid water as heat energy is absorbed by the cold surface. The effect is called condensing. But again, this process takes time.

Solid, liquid, and gas are the three common states (or phases) of matter that we encounter every day. Melting, freezing, condensing, and boiling are called changes of state (or phase). Like specific heat capacity, these are quantified physical properties that involve heat energy. Now that we understand more about the nature of heat and temperature, we can take a closer look at these physical changes.

Ponds do not freeze immediately when the weather falls below 0 °C. The snow on the ground and the frost on the trees show that the temperature of the air is clearly below freezing, but the pond water is still liquid. The water needs time to release a lot of heat energy as it turns to ice.

Experiment: Going Through a Phase

When enough heat energy is added or removed from many substances, a phase change occurs. This activity observes the temperature change as ice turns to water. **Before starting, read through the experiment and create a data table.**

Materials: plastic water bottle   water   hot plate
              large beaker or saucepan   stopwatch   thermometer

Procedure: Fill the plastic bottle about halfway with water. Place the thermometer in the water as shown in the picture below. Place this assembly carefully in a freezer, and leave it undisturbed for several hours or overnight, until it freezes solid.

Turn the hot plate on. Pour a few centimeters of water into the beaker or saucepan, and place the plastic bottle in. Measure and record the starting temperature, and start the stopwatch. Place the beaker on the hot plate. Measure and record the temperature every minute until the temperature reaches 20 °C or more. Stir the ice/water with the thermometer between measurements when the ice melts enough to do so.

Observe the water during the experiment, and record any changes that you see and the stopwatch time that the change occurs.

Data: Make a data table before starting the lab. Start with lines for 12 minutes of data, but leave enough room to extend the tables if needed.

Analysis: Construct a graph showing temperature on the y-axis and time on the x-axis. Sketch in a curve that follows the data points.

1) What significant physical change occurred to the water? At what point (time and temperature) did that change occur?

2) Interpret your graph and comment on the rate of temperature changes measured. Did the temperature change at a constant rate, or did the rate change during the experiment?

3) Was the temperature constant (or nearly constant) during any time period? When?

4) During what time period was heat being added?

5) Consider your answers to the questions above, and the shape of your graph. Describe what was happening during the period of constant temperature.

6) Create a graph that shows an approximate temperature/time curve for a piece of ice in a bowl on a hot plate. The ice starts at -12 °C, warms, melts, and continues to warm until the water boils. Assume that the entire experiment happens in about 20 minutes. Explain what is happening to the sample during each section of the graph.

2.5 Latent Heat

Careful experiments can sometimes reveal puzzling details of a physical process that cannot be observed with the eye. Experiments reveal that the temperature of the substance does not change during the melting or boiling process. The sample continues to absorb heat, but the temperature remains constant as the relationship between molecules changes. While a substance undergoes a phase change, it remains at constant temperature while the phase change occurs.

When a substance melts, the structured relationship holding molecules together as a solid relaxes, allowing the molecules to move freely about. The energy needed to separate the solid is called the **latent heat of fusion**. The Latin word "latent" means *hidden*. In this case, the heat energy is hidden because it is not measurable as a temperature change. The term "fusion" is also of Latin origin, and it means *melt*. The word fusion is also used in other ways, particularly when two things are combined, such as a fusion of two types of music, or when food from different cultures is combined.

When the substance is cooling from liquid to solid, the molecules form a structure and hold their place within the substance. The term for this reverse process is **latent heat of crystallization**, but the amount of energy needed for the change is the same in either direction. An ice cube put into a warm beverage will melt slowly, absorbing heat from the liquid as it changes phase. Removal of heat from the liquid cools your drink.

When a substance begins to change phase from liquid to gas, the fastest-moving particles have the most kinetic energy. These particles are the most likely to escape into a gas phase, leaving the slower-moving particles behind. The remaining liquid continues to absorb energy. The temperature does not increase because the fastest-moving particles escape. The energy needed for this process is called **latent heat of vaporization**. In the reverse process, gas particles slow down and revert to the liquid phase. This loss of energy reveals the **latent heat of condensation**.

LATENT HEAT OF FUSION is the amount of heat energy needed for the change from solid to liquid. The reverse process is called CRYSTALLIZATION.

LATENT HEAT OF VAPORIZATION is the amount of heat energy needed for the change from liquid to gas. The reverse process is called CONDENSATION.

Like specific heat, latent heat is a property that is usually specified as energy per gram. However, the units for latent heat are J/g (joules per gram), while the units for specific heat are J/g•°C. Since the temperature remains constant during the phase change, there is no need to multiply by a change in temperature. The calculation for the amount of heat needed for a phase change is simply:

$$\Delta H = mc_f \quad \text{or} \quad \Delta H = mc_v$$

The first calculation is for a melting or freezing process and the second is for a boiling or condensing process.

The subscripts in c_f and c_v simply indicate heat of fusion and heat of vaporization. The term c_f can be used for fusion or crystallization, and c_v can be used for vaporization or condensation, because the amount of energy needed is the same in either direction of the phase change.

In some cases a substance can change directly from a solid to a gas, skipping the liquid phase. One notable example is carbon dioxide, known as dry ice in its solid form. At atmospheric pressure, dry ice **sublimes** – it changes phase directly from solid to gas. The liquid phase of CO_2 can only be attained at extremely high pressure.

SUBLIME: to change phase directly from solid to gas. Many substances sublime under the right conditions.

A lot of energy is needed for a phase change, but the heat from this volcano can do the job. Most of what looks like smoke coming from the volcano is actually water vapor, which becomes visible as it hits cooler air and condenses into tiny droplets of liquid water.

Many substances can sublime under just the right conditions, although the process usually takes a long time. Regular water ice sublimes very slowly in your freezer. Some freezers are advertised as "frost-free". These devices have a fan that keeps air inside the freezer moving, which increases the sublimation process of any frost that might otherwise build up in the freezer. If you have a "frost-free" freezer, check the ice cube trays; after a week or two, the cubes will noticeably shrink as the ice sublimes!

Recent analysis of photographic images of the polar ice caps on Mars confirms that common water ice sublimes rapidly in the thin atmosphere. Solid water also sublimes in the vacuum of space, and this process was used to cool the spacesuits of astronauts on the moon. Some metals, such as zinc and cadmium, will sublime in very low pressure conditions, so engineers must be cautious about using these metals in spacecraft. Sublimation, and the reverse process, called deposition, are very interesting phase changes. The heat exchange is the sum of the heat of fusion and the heat of vaporization, though there are some complications to this calculation. Further discussion of this topic is not necessary here, as we will restrict our study to melting and vaporizing.

When a substance burns, it is a *chemical* change; but a phase change is a *physical* change. Certainly the substance is altered somewhat; after all, the properties of ice are different from those of water. However, any phase change is easily undone by the opposite process, which returns the substance to its original state. The reversibility classifies phase and latent heat as physical properties.

Extending Ideas

(use the latent heat table below to help solve these problems)

1) How much heat is needed to melt 30.0 grams of ice, if the ice starts at 0.0 °C?

2) How much heat is required to change 20.0 g of ice at -10.0 °C into water at 22.0 °C? (hint: do this problem in three steps)

3) How much heat is required to turn 18.0 g of ice at -12.0 °C into steam? Remember to account for *two* phase changes!

Latent Heat of Some Common Materials

Material	Heat of fusion	Heat of vaporization
Aluminum	396 J/g	10,500 J/g
Copper	205 J/g	5,390 J/g
Ethyl alcohol	104 J/g	858 J/g
Gold	64.5 J/g	1,930 J/g
Iron	267 J/g	7,230 J/g
Lead	24.7 J/g	944 J/g
Mercury	12.1 J/g	324 J/g
Nickel	300 J/g	7002 J/g
Silver	105 J/g	2,700 J/g
Water	334 J/g	2,256 J/g
Zinc	102 J/g	2,000 J/g

3
Heat Flow and Efficiency
3.1 Thermodynamics

Water has figured prominently in our study of heat energy. Several factors make this substance very useful in studying heat: water is very common, inexpensive, chemically stable, and has a high specific heat capacity. These qualities also make water extremely important in the heat exchange processes that control climate and weather patterns on earth.

Water is also used extensively to store and transfer energy in many of the machines that make our lives easier. Any of these processes, whether natural or man-made, require exchange of heat energy between water and some other substance.

Of course, weather and climate affected our earth for eons before we understood anything about heat energy. As our world becomes ever more crowded, the need to understand and predict natural forces grows more important. Clever engineers and inventors created engines and heating devices before scientists had a clear understanding of the science behind how they functioned. Today, the increasing need for efficient use of energy requires sophisticated analysis of heat flow and energy distribution.

Heat flow between air, water, and land heavily influence weather and climate. Sunlight, earth's rotation, and the tilt of earth's axis are other major factors.

When heat is gained or lost, the molecules of a material show that change with an increase or decrease of motion. The energy of motion (kinetic energy) of these particles is often observed as a change in temperature or phase. Temperature changes can be measured with a thermometer, but phase changes involve a gain or loss of heat energy without a temperature change. The particles of a solid release their structure and liquefy. The particles of a liquid escape from their neighbors and become a gas. These processes were observed in your lab activities in the previous chapter. A time vs. temperature curve and the calculation of latent heat provide some insight to phase changes.

This chapter will look at some observable changes that can occur when heat energy is gained or lost. The study of energy flow and the effects of adding or removing heat is called **thermodynamics**.

Knowledge of this subject will help you improve the efficiency of everyday processes in your home, reducing your family's energy consumption. Helping to solve global energy issues begins with your own little corner of the world!

THERMODYNAMICS is the study of thermal energy, heat transfer, and applications of this knowledge.

We will continue our study of thermodynamics with a look at the way heat flows within substances and the ways that it travels from one substance to another. All of this talk of heat flow might remind you that scientists once thought of heat as a fluid that can move between objects. Despite these comparisons, remember that heat is not matter – it has no mass, and is not made of atoms. Very precise experiments show that adding heat energy to a substance does not increase the mass. Furthermore, matter occupies space, so two different substances cannot be in the same place at the same time. Heat can move right through matter, as you know from many common experiences. This is more confirmation that heat is not a substance, but rather a type of energy that can be changed into other forms of energy.

Drops of water often condense on the outside of a cold can or glass in warm humid air. Water vapor dissolved in air loses energy when molecules strike the cold glass. Loss of energy from gaseous water changes its phase into a liquid.

Experiment: Chill Out, Man!

This activity explores the cooling rate of samples of water in different containers. We will graph the data and consider the factors that affect the cooling rate.

Materials: flask   metal can
                 glass beaker   foam cup
                 paper cup   thermometer
                 hot plate   heat-resistant gloves
                 OPTIONAL: basin containing a "water bath"

Procedure: Pour water into each of the four containers – metal can, glass beaker, foam cup, and paper cup – to a depth of about 2 to 3 cm. This is just to make sure that you have enough water. Then pour all of this water into the flask (it should not be more than about 2/3 full).

Put the flask on the hot plate and set the hot plate to a medium temperature. While you are waiting for the water to heat, measure the temperature of the room, and create a data table as shown below.

Heat the water in the flask until it is between 40 and 50 °C. This

Time:	0 min	1 min	2 min	3 min	4 min	5 min	6 min	7 min	8 min
metal									
glass									
paper									
foam									

is not hot enough to burn you, but hot enough so that you should avoid spilling it on you. When the water temperature is within the required range, turn off the hot plate and record the temperature in your table as the starting temperature for each of the containers, at time "0 minutes". Use your table to record this and all other temperature measurements.

Start the stopwatch and pour water to a depth of about 2 to 3 cm into each of your containers. If you do not have heat-resistant gloves, you may grasp the flask with your hand <u>by the neck or handle</u> when pouring. If you hold the flask below the water line, it will be too hot to hold. The amount of water in each container should be about the same, but this is less important than the need to do this efficiently, so that each container has about the same starting temperature (in other words, pour the water as quickly as you can safely do it).

Repeat the temperature measurements in the same order every minute. Begin with the first container that you filled, stirring with the thermometer for about 20 seconds before each reading.

Data: Obtain measurements for 8 minutes to fill your table. Record the room temperature (or bath temperature if used) and label it clearly.

Analysis: Make a single graph for all of your data; time should be on the horizontal axis and temperature on the vertical axis. Graph the data from your table; be sure that the points for different containers are marked differently (different symbol or color or both). Join the points with an approximate curve. Use a pencil for the curve so that you can sketch and correct until it looks like a smooth curve.

1) List the containers in order from best insulator to worst.

2) How did you determine the order for the list?

3) Assuming that each sample contained 50 ml of water, calculate the heat energy that each lost during the entire time period measured. (Remember: $\Delta H = m\,c\,\Delta T$)

4) Which sample lost the most heat? Did this sample lose the same amount of heat energy during the first minute as it did during the final minute?

5) How can you compare the rate of heat energy lost (the cooling rate) for the samples just by looking at the graph?

6) If you were to continue all of the curves for the rest of the day, what is the lowest temperature they would eventually reach, and why?

7) What factors do you think affected the rate of cooling? (Hints: Why did the samples cool at different rates? Why did the cooling rates change as time went on?)

8) What would happen to the temperature of a very hot pitcher of water left in a 20°C room overnight? What is the lowest temperature it would reach, and why do you think so?

3.2 Thermal Expansion

"Heat rises" is an old saying that appears to be a scientific law. Like many old sayings, this one is based on common observations and therefore has some claim to truth. However, this does not make it a law, and you can quite easily show a counter-example. Put one end of a metal spoon in a flame, and hold the other end lower than the flame. The end you are holding will soon be too hot to handle! Heat moved *down* the metal rod, from the flame to your hand. If we look carefully, we will find the science behind the old saying and reach a better understanding.

Adding heat to any substance causes an increase of the motion of atoms. This faster motion requires more space, whether the substance is a solid, liquid, or gas. Therefore, a substance takes more volume when it is hot than when it is cold. If you have ever seen people dancing, you can draw a comparison. When the tempo of the music increases, people dance more energetically. They spread out on the dance floor to make more room for their additional movement.

A heating coil held over the top of a chunk of metal. As you can see by the glow, heat moves *down* through the metal to the lower end. Heat does not always rise!

> Nearly all substances occupy more volume when heated. This is known as THERMAL EXPANSION.

The principle applies to atoms as well as to dancers. More energy means more motion, and more motion requires more room to move. In general, adding heat energy increases volume, a process known as **thermal expansion**.

Thermal expansion of a gas is especially noteworthy. The lower density of hot gases causes them to rise above cooler, denser gases. Hot air balloons work this way, lifting not just the air within, but the balloon itself and the basket hanging below. In a car engine, the heat of burning gasoline causes trapped air to expand in a small cylinder. The hot air pushes a piston, which drives a rod attached to a crank and makes the car move. Those expanding gases can do a lot of work!

Thermal expansion affects liquids as well. Warm water rises to the surface of cold water, and hot oil rises to the surface of cold oil. But even very hot water is denser than oil, so cold oil will float on hot water. *Heat* doesn't rise, but a hot fluid is less dense than a cold sample of the same fluid, and will rise to the top.

The expansion of fluids also explains how a liquid thermometer works. When the thermometer bulb is in contact with a warmer or colder substance, the liquid inside (usually colored alcohol or mercury) increases or decreases in temperature, and the reservoir of fluid contained in the bulb expands or contracts. The liquid inside the thin tube indicates the temperature on the scale. Of course, the thermometer liquid must absorb heat from the sample, which is why you must allow some time for the thermometer to reach the same temperature as the surroundings. This also means that when you use a thermometer, you change the temperature of the substance you are measuring just a little bit!

In solids and liquids, the rate of expansion is very small, but the effects can be quite interesting. When ordinary glass is heated or cooled unevenly, the expansion is different from one side to the other, sometimes causing the glass to crack. Special glass with very low thermal expansion is used in cooking or in the science lab to prevent such mishaps. Bridges are built with expansion joints that allow room for the road surface and supports to grow and shrink with the change of seasons. Without these gaps in the structure, the bridge could buckle in very hot weather. The familiar clickety-clack of a train comes from the wheels traveling over small gaps in the track, left between rails for the same reason.

This bridge expansion joint provides a space that allows the bridge to grow longer in warm weather, but cars can roll smoothly over the gap.

You have probably heard that scientists are concerned that global warning will make sea levels rise around the world. One reason for this rise is the melting of ice sheets over Greenland and the South Pole. However, the increase in temperature will also cause expansion of water in the oceans, contributing to the increase in sea level. Although the rate of expansion of solids and liquids is quite small, the effects can be very significant!

There are a couple of interesting exceptions to the typical thermal expansion of substances. The most important exception is ordinary water near its freezing point. This unusual property is discussed more in section 3.4.

Air is heated by a flame in the lower end of a hot air balloon. The air expands, and some escapes from the bottom, leaving hot, low-density air in the balloon. Even with the basket and people inside, heater, ropes, and the balloon itself, the entire contraption is less dense than the cool air around it, allowing the balloon to rise.

Practice Problems

1) A thermometer can be described as a device that only measures the temperature of itself. Do you agree with this description? Write a paragraph that either supports or argues against this definition.

2) What properties are needed for the liquid used in a thermometer? For each quantifiable property that you name, indicate whether the desired value should be a high or low number.

3) Your home is maintained at a comfortable temperature year-round, although sometimes it may be too warm or too cool for your liking.
   a) Name three sources of energy entering your home.
   b) Name three ways that energy leaves your home.

4) Your body is maintained at the right temperature by your metabolism. What is the source of this energy, and how does your body reduce temperature when you work hard?

5) Holding a snowball causes your hand to feel cold. What happens to the heat? What is the effect on the snowball?

6) Research: A car engine can overheat because the energy released by burning fuel is not all transformed into useful work. How is excess heat energy removed from a car engine?

Extending Ideas

Use your graphs and results from the "Chill Out" experiment to guide your answers to these questions

1) Imagine the following experiment:

   A cup of cold water is placed in a pot of very hot water (80 °C); another cup of cold water is left in the open air (20 °C). The initial temperature of each cup of cold water is 4 °C, and the temperature is measured each minute for the next twenty minutes. On one graph, show the change in temperature you expect for each cup.

2) Imagine the following experiment:

   Three cups of cold water (4 °C) are left in the open air (20 °C). The temperature is measured each minute for the next twenty minutes. One cup is made of paper, one of glass, and one of metal. On one graph, show the change in temperature you expect for each cup.

3) "A watched pot never boils" is an old saying. Why does it seem that a pot of water warms up very quickly, but takes a very long time to actually begin boiling? Use your knowledge of heat capacity, heat transfer, and latent heat to answer this question.

3.3 Heat Transfer

In our experiments, heat energy always moves from the sample at a higher temperature to the sample at a lower temperature. As time passes, the samples reach thermal equilibrium; they are at the same temperature. How does that energy exchange occur, and what factors affect the rate of this process? We shall now explore these ideas more deeply, drawing on your everyday experience to describe how heat energy moves around.

If you heat a bowl of soup on the stove, the burner applies heat to the bottom of the pot. The soup at the bottom gets hot first, expands a bit, and begins to rise. Cooler soup from the top of the bowl sinks to replace the rising soup, and the cool soup is now nearer to the burner. The process continues and the entire pot of soup gets hot. This process is known as **convection**, and is chiefly responsible for distributing heat throughout most fluids.

CONVECTION distributes heat throughout fluids. Fast-moving molecules carry energy to other molecules in the fluid.

Convection is the dominant type of heat transfer in a fluid. Air warmed by the hand spreads the energy to other molecules. A special type of photography captured this image.

If a fluid is not free to move around, the heat energy can be trapped. This is how double-pane windows keep heat energy inside your home. Trapped air between layers of glass slows down convection. This is also why wearing layers of clothes or loose knits can help keep you warm in the winter. Air becomes trapped within the clothes, and contains your body's heat energy. It isn't the cloth that keeps you warm, but rather the air that is trapped *between* and *within* layers!

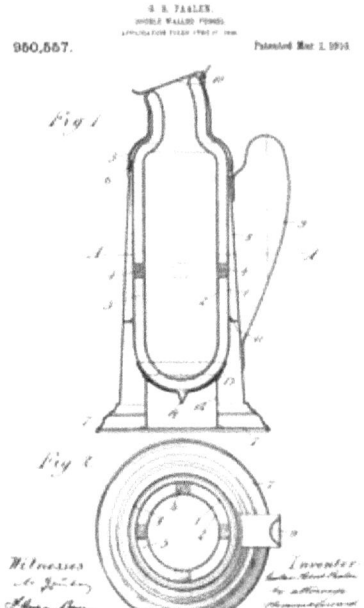

An excellent heat container is made from a double-walled container, as shown in this drawing for a patent application. Trapped air between the walls reduces convection. Pump the air out (a vacuum) to eliminate convection completely.

Convection cannot take place in a solid. The atoms of a solid are not free to flow, so the hot molecules cannot carry their kinetic energy to other places in the substance. Nevertheless, heat travels from the burner on the stove through the pot to heat up the soup. Heat passes through the walls of each container in your previous experiment. Heat moves through the glass of a thermometer to affect the liquid inside. How does heat energy transfer through solids?

Stretch a rubber band tight, and pluck it like a guitar string. You can see and hear the vibrations. You sense this vibration energy as sound. When you tap the end of a metal rod, the entire rod vibrates. If the rod is very thin, you can see this motion. If the rod is thick, the motion is too small to see – but you might hear a tone that results from these tiny vibrations. The energy from your tap travels through the material, setting the entire rod in motion. The vibration energy also transfers from the rod to the air.

On an even smaller scale, heat increases the vibrating motion of atoms. Atoms pass vibration energy to nearby atoms; this is how heat energy can travel through solids. These vibrations can also transfer to other substances that are in direct contact with the solid. Within a material or from one substance to another, direct heat transfer between atoms is called **conduction**. The ability of materials to conduct heat varies widely. Most metals are excellent conductors, a property that can be explained by the metallic bonds that hold the nucleus of metal atoms in a sea of electrons. Most ionic compounds and covalent compounds are poor conductors of heat. Substances that are poor conductors are called **insulators**.

A frying pan must be made of a good conductor – iron or aluminum are commonly used. The handle must be a good insulator, or the chef must use an insulating pad to avoid a painful heat transfer lesson!

Heat can move through solids by passing along molecular and atomic vibration. This type of heat transfer is called CONDUCTION.

INSULATORS reduce heat transfer. Good insulators do not transfer molecular vibration very efficiently.

Some common insulators are ceramic, wood, and rubber. Many utensils in the kitchen have handles made of one of these substances. Trapped fluids can also be excellent insulators, as described earlier. However, the best insulator of all is *nothing*. If the air between layers is pumped out, there is no possibility for any conduction or convection, so heat can be very well contained. A space with nothing in it – not even air – is called a *vacuum*.

RADIATION energy travels without any molecules to transfer vibration. When the radiation hits a substance, energy is absorbed as heat.

There is one more important type of heat transfer: **radiation** describes the way energy can travel through space and excite motion in atoms that are far from the source. Energy from the sun reaches us through radiant heat transfer. All objects emit some amount of radiant energy, but only those that are hotter than the surroundings emit more energy than they absorb. Objects that actually glow release a lot more radiant energy than they absorb. Fires, light bulbs, and heating elements on electric stoves are all sources of radiant heat.

Old-style light bulbs turn most of the electrical energy into radiant heat, and only a small amount into light.

Do not confuse radiant heat energy with radioactivity. When a nucleus breaks down and releases energy, particles are emitted along with very high-energy radiation, and exposure is dangerous. Radiant heat occurs at much lower intensity and does not include the nuclear particles that are emitted as part of radioactive decay. Nevertheless, high-energy heat radiation is dangerous, which is why you must shield yourself from excessive exposure to sunlight, and you can be burned if you sit too close to a household radiant heater.

Several ideas in this text describe a set of physical laws called the **Laws of Thermodynamics**. Conservation of Energy, described in Chapter 1.3 and used throughout the text, is also called the **First Law of Thermodynamics**. The fact that energy spreads out (dissipates) as time passes is described in Chapter 2 and called *entropy*. This tendency toward less order is the **Second Law of Thermodynamics**. The impossibility of reaching the lowest theoretical temperature of "absolute zero," is called the **Third Law of Thermodynamics**. As described in section 1.8, this law states that it is not possible to remove the last bit of molecular kinetic energy to reach zero K.

Scientists then realized they needed another law that was even more basic. This newly recognized law was called the **Zeroth Law of Thermodynamics**. It states that two substances that are each in thermal equilibrium with a third substance must also be in equilibrium with one another. This explains how we know that two substances reach thermal equilibrium by measuring their temperature (the thermometer is the "third substance").

3.4 Energy and the Earth

93 million miles (150 million kilometers) from earth, the sun creates energy from matter by a process called nuclear fusion. Earth receives some of this energy across that vast distance. How does that energy get here, and what happens then? Thermodynamics affects your daily life, as this energy is transferred around the globe. The three main heat transfer processes are responsible, each contributing to the flow of energy.

RADIATION is the means that energy from the sun can travel to the earth. Space is almost a complete vacuum – 150 million kilometers of nothing between the sun and the earth. However, the sun releases a tremendous amount of energy, and the earth is relatively small; so only a tiny percentage of that energy reaches earth.

The oceans absorb part of the energy, and land absorbs a smaller amount. CONDUCTION from the surface carries heat to the air above and to the water below. The uneven heating and the spin of the earth create ocean and wind currents, and CONVECTION distributes this energy around the globe and to greater heights and depths. The shape and location of continents and geographical features (such as mountains and valleys) influence the movement of both water and air. The flow of heat energy through these two huge fluid reservoirs creates weather patterns.

The round earth does not receive this energy evenly over its surface. Furthermore, the earth spins on a tilted axis, which means that sunlight reaches the atmosphere at an angle that changes as the earth circles the sun. During the summer, the northern half of the earth is tilted toward the sun. We get more hours of sunlight, and at a more direct angle. During the winter, we are tilted away from the sun – fewer hours of light, and at a glancing angle.

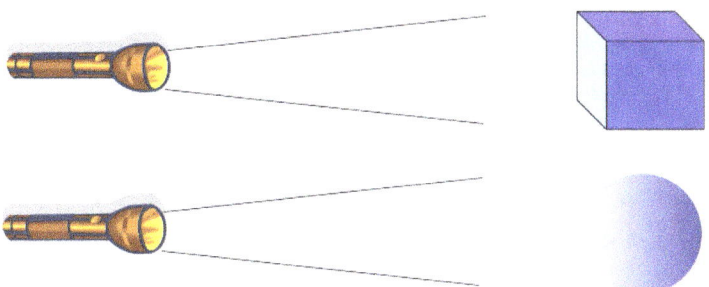

Shine a light on a cube, and only the side facing the light is illuminated. The rest of the cube is dark. Shine a light on a sphere, and the reduction of light is gradual as the angle between the light and the surface of the cube increases. This is why the earth's equator receives more direct sunlight than the poles. Throw in a spin and we have day and night; add a tilt to the axis, and we have seasons!

Convection is a significant process in the oceans and the atmosphere. Convection is messy, turbulent, and irregular. Clouds, snow on the ground, and other seasonal changes can change the amount of heat absorbed, adding another level of uncertainty to weather predictions. Even the most sophisticated computers can only come up with a probability of future weather conditions. You may find this especially frustrating when the chance of rain is 50%. With all of the satellite pictures, instrument readings, and supercomputers, the best they can say is: "maybe it will rain…and maybe not."

Weather is an extremely complex thermodynamic process, and it seems that the only reliable factor is chaos. Very similar starting conditions can end with completely different results. However, our study of heat reveals some interesting factors that we can count on.

Water can absorb, store, transport, and release large amounts of heat energy. The high heat capacity of water - and the amount of the earth that it covers - means that bodies of water have a great influence on weather and temperature. These effects can be on a small scale, like a cool breeze off the lake; or a huge scale, like a devastating hurricane that gains energy over a warm ocean. The outcome can last for minutes, like a summer shower, or for all known history and as far as we can predict into the future, like the Gulf Stream.

Large bodies of water have a moderating effect on temperature swings. In winter, it is often warmer on the coast than just a few miles inland, and in summer, the reverse is true. On the hottest summer day, the water in a lake remains relatively cool, absorbing heat from the surrounding air.

The high heat capacity of water stores energy in winter, and absorbs energy in summer. The result is moderate temperatures near the coast. When warm, moist air over the water meets cooler air over the land, heavy fog or precipitation occurs.

Another unique property of water explains how fish can survive the winter in a pond. Most of the time water contracts (shrinks) when the temperature drops, just like other substances. But as its temperature decreases from 4 °C to 0 °C, water actually expands as tiny ice crystals begin to form. These crystals do not fit together as tightly as liquid water molecules. For most other substances, the solid form is denser than the liquid, but water is the opposite.

One effect of this is that the very coldest water in a lake or pond rises to the surface, where it is either warmed by the sun or freezes. If it freezes, the ice helps insulate the water, reducing the loss of heat due to convection from the water below. The water under the ice remains liquid throughout the winter, and fish can survive until spring!

Extending Ideas

1) Section 3.3 makes the following statement about radiation: "All objects emit some amount of radiant energy, but only those that are hotter than the surroundings emit more energy than they absorb."
    a) What does this statement imply about objects that are cooler than their surroundings?
    b) What does this statement imply about objects that are the same temperature as their surroundings? What is the word for such a situation?

2) In section 3.3 is a picture of an old-style (incandescent) bulb. The caption states that this kind of bulb converts most of the electrical energy to radiant heat. Research different kinds of light bulbs, and find out which are best at converting electrical energy to light instead of wasting it as heat.

3) Watch the weather report on the evening news. Write down the qualitative and quantitative descriptions used for tomorrow's weather prediction.

4) Copy or clip and paste a weather map that shows temperature measurements for the entire country. Compare these measurements, and comment on the conditions near each measurement that affect the local temperature.

5) Find a local newspaper with a weather forecast for all of next week. Clip the forecast and tape it to a piece of paper, or copy down the information for each day. Record the actual weather conditions as each day passes. At the end of the week, compare the actual and predicted weather. Comment on the accuracy of the predictions as the days pass since the original one-week forecast.

3.5 Heating and Cooling Systems

People are only comfortable in a small range of temperatures, while the range of weather found on earth varies widely. Long ago in human history, this fact restricted the spread of population to the warmer regions of the earth. Human populations often traveled to follow the comfortable weather and available food. The development of agriculture and basic architecture led to permanent communities. Changing seasons and variations in weather made it necessary to develop ways to control the temperature in homes, whether those homes were caves, huts, houses, or castles. There are many different kinds of heating and cooling systems in use throughout the world, but all have several features in common.

Today, all homes are designed with some kind of heating or cooling system, or both. Heating systems in some parts of the world consist of centrally located open fires, a simple but effective technique used for thousands of years. Among wealthier populations, heating and cooling systems tend to be much more elaborate. These methods are required to moderate the temperature throughout much larger homes, often in a wide variety of climates. Despite the differences, there are some similarities between those heating systems that were available many centuries ago and those that have been devised only recently.

All heating systems require an **energy source** of some kind. Most often, a fuel is burned, changing chemical energy to heat energy. Common fuels include coal, oil, natural gas, and wood. Many other organic materials are used around the world, depending on what is locally available. The chemical reaction of burning produces gases that

Most home heating systems burn fossil fuels as the main ENERGY SOURCE. If electric heating is used, the energy is often produced from fossil fuels. So the energy used to heat most homes comes from fossil fuels.

A WORKING FLUID absorbs thermal energy at the heat source and moves it around the home.

HEAT EXCHANGERS aid the energy transfer into and out of the working fluid.

must be ventilated from the building. Unfortunately, a large portion of the heat energy is lost along with these gases.

Other energy sources include solar or geothermal heat. These take energy directly from the sun's rays or from heat sources within the earth. Also, electrical energy can be turned directly into radiant heat energy. This energy conversion works just like in a toaster, but on a larger scale. This method is common in places that do not get very cold for most of the year.

Heat energy must be distributed around the house, so a **working fluid** is required. This substance absorbs heat from the energy source and then circulates around the building. Heat energy is carried by the working fluid to all parts of the house.

Of course, the burning fuel must be controlled and contained to protect the occupants of the building and produce the right amount of heat. A **heat exchanger** allows energy to move from the source to the working fluid, or from the working fluid to the air in the home. Heat transfer can occur through metal pipes, the brick surrounding a fireplace, the surface of a terra-cotta oven or metal stove, or through ducts within a furnace. Some energy sources – such as fires and electric radiators – transfer most of their heat through radiation without passing through another material.

Air is quite commonly used as a working fluid. This makes a very simple system, because the air simply drifts past the heat transfer surface or is forced through a duct by a fan. The air then spreads throughout the building. Water is an alternative working fluid, but of course it must be sent to and from the energy source in pipes. Water is sometimes vaporized to steam by the energy source, then returns to liquid as it releases heat to the air. The high latent heat of condensation means that a little steam can transport a lot of energy. Transferring heat from the water or steam to the air in the building requires another heat exchanger, such as the radiators shown here.

Hot water is the working fluid in many homes. It is piped to different rooms, where a radiator helps transfer energy from the water to the air. Radiators like this are heat exchangers.

Finally, energy in the building must be contained by **insulation** on all exterior surfaces. Hot air must be kept in the house, and heat loss through the walls and windows must be minimized. Various poor conductors are used for this purpose, often using trapped air to help prevent heat exchange.

INSULATION slows the transfer of heat energy.

Cooling systems may require many of the same components as heating systems.

Cooling systems require much of the same kind of equipment; after all, the goal is similar. In both systems, heat energy is moved from one place to another. The simplest cooling methods rely on exchange of cooler outside air for warmer air trapped in the house. Cooling systems may be little more than well-designed ventilation methods that were developed over many centuries of trial-and error. Air exchange can occur without any extra machinery, but a fan can aid the process.

Air conditioners are not the only kind of cooling system – and they are certainly not the most energy efficient! Ventilation and water evaporation are often used to cool homes in hot climates.

The WORKING FLUID is a gas or liquid that carries heat energy in a heating or cooling device. It may or may not change phase as it does so.

Some cooling methods rely on the phase change of a **working fluid**. Many traditional cooling systems in hot climates use evaporation of water from a surface that is exposed to the air flow. You could describe sweat as the working fluid that cools you when a fan blows on your face.

Air conditioners use a working fluid that is trapped within the device. Pumps circulate the fluid, which absorbs heat energy from air in the room through a metal duct with fins protruding from the surface. This device is the heat exchanger, and it works just like a radiator in reverse. Heat energy travels *from* the air and *into* the working fluid, which changes phase from a liquid to a gas. Compressors squeeze the working fluid back to a liquid, releasing energy to the outside of the building. Again, good insulation is necessary to keep heat energy on the other side of the door!

Evaporation is a change of phase that happens at the surface of liquids. Particles at the surface can escape at temperatures below the boiling point when the fastest-moving particles escape, leaving the slower ones behind. Remember, slower means cooler – so evaporation cools the surface when this energy leaves. Sweat is your body's air-conditioning system. When sweat evaporates from your skin, you cool down. You must remember to replace that liquid in your body by drinking water!

Extending Ideas

1) The previous section describes the essential parts of home heating and cooling systems. Investigate the method used in your home. What is the energy source? The working fluid? How is energy delivered to different rooms? Is there more than one source of heat energy? What are the ways that heat leaves your home?

2) How is your home cooled on hot days? How much of your home is cooled by each method? Think of three ways that the energy consumption needed to cool your home can be reduced.

3) A piece of metal and a piece of wood that are in the same room should be at the same temperature. Find one of each, and feel them with your hand. Does one feel cooler than the other? Explain how the action of conduction could affect how two objects at the same temperature could feel different.

Ice cubes float in a drink (which is mostly water), showing they are less dense than the liquid. They will melt slowly, absorbing heat from the liquid as they change phase. The latent heat of fusion is responsible for your cool, refreshing beverage.

3.6 Engines and Power Plants

Many of the complex mechanical devices that we use daily draw energy from the expansion of a gas when heat is added. We have already described the moving pistons in engines of cars and other motor vehicles. These vehicles nearly always burn gasoline or another fuel derived from oil. Jet airplane engines also use a type of refined oil, but instead of pushing on a piston, the expanding gas rushes past a kind of pinwheel. The technical name for the pinwheel is a *turbine*; it is attached to a fan that draws in more air to continue burning fuel. The expanding gas continues out the back of the jet engine, propelling the airplane just like the air rushing out the back of a balloon. Hopefully, the airplane is under a little more control than the balloon!

Most electrical energy in the United States is produced in large power plants. Chemical energy from coal, oil, or natural gas provides energy for many of these power plants. Heat from nuclear breakdown is also a common energy source. In either case, the heat energy boils

The steam locomotive and the steamboat were two forms of mass transit that became practical when engineers learned to harness heat energy.

Today, ships and trains are powered by much more efficient engines. The more we understand about heat and energy changes, the better we can use that knowledge to improve our world!

water, and this expanding water vapor rushes past turbines, just as in a jet engine, but much larger. These spinning turbines run electrical generators that deliver electricity to the area through a network of wires and transformers.

All of these devices consume fuel and produce work. Heat energy always disperses as a by-product of this process. Energy changes from a concentrated, stored form and becomes spread out and unusable. As the desired work is performed, even more energy ends up as dispersed heat. Later, we will learn how friction transforms work into heat. In all of these examples, the energy used spreads out to a more uniform distribution. The universe tends toward overall equilibrium, and the energy that we use accelerates this process.

Activity: Efficiency

Hot baths and showers, washing dishes, and cooking all require hot water. If water is the working fluid in your home heating system, most of the energy used in your home is directly applied to heating water, consuming most of the total energy used in your home.

There are many ways to heat water, and they are not equally efficient. This activity explores the efficiency of several options available in the kitchen.

Materials: water   thermometer   digital balance
              beaker   various electrical heating devices

Procedure: Your teacher may ask you to test one or more different heating devices. For each test, measure the starting temperature and the amount of water. Instructions for each device will tell you the energy used per second and the heating instructions. If the water boils during any heating process, you must find the exact amount of water remaining, and include the heat of vaporization when you calculate the energy absorbed by the water.

Data: Energy use is measured in joules per second (J/s). Record this information for each heating device. Use the electronic balance to measure the amount of water that you start with in each case. You will also need to measure the water after heating, to find out how much (if any) has turned into vapor. You will need the starting temperature of the water sample and the final temperature after heating. Record the heating time and the name of each device.

Analysis: For each heating device that you tested:
1) Multiply the joules per second times the time to find the energy consumed by the device.

2) Measure the amount of water remaining, and calculate the amount that has boiled away. Use the formula $\Delta H = m \, c_V$ to calculate the energy absorbed during the phase change. Remember to use only the mass of water that boiled away.

3) Use the formula $\Delta H = m \, c \, \Delta T$ to calculate the energy absorbed during the heating process. Use the mass of *all* of the water here.

4) The efficiency is equal to the percent of energy used that actually accomplished the required task. In this case, divide the energy absorbed by the energy consumed by the heating device.

5) Describe and compare the efficiency of the devices that you used, along with any ideas that might increase the efficiency of heat transfer.

3.7 Consumption and Conservation

The science and technology described in this chapter is fairly recent in terms of human history. Elaborate home heating and cooling systems, power plants that produce and deliver electricity, and machines for manufacturing and transportation were not around when the country was founded. These inventions make our lives easier, and give us the opportunity to devote more time to learning, creating, and enjoying life rather than spending each day merely trying to survive. Of course, there are consequences as well as benefits to this progress.

Scientists have learned a lot about heat and energy in the past two hundred years. Our knowledge of energy transformations and heat transfer allows us to live comfortably throughout the year in buildings that are warmed or cooled to our liking. Thermodynamics powers the engines of ships, trains, automobiles, and airplanes. All power plants create electrical energy through some process of energy change, usually involving heat, phase changes, and thermal expansion.

Typically, these complex systems require a relatively large investment of both original cost and continuing use of resources. Many details can influence the efficiency of these devices. Some are adapted for specific building types or typical weather conditions. More expense and effort must be put into the heating system of a house in Canada than one in Mexico. Foresight and thoughtful design can reduce the expense of maintaining a comfortable temperature inside houses and office buildings. Methods of heating and cooling have been refined in the last half-century, as growing population and diminishing resources raised increasing concerns that conservation is necessary.

All life requires energy to sustain itself. For most living things, that energy is obtained from the sun. This energy travels through space to reach earth; we will study how that happens in the next chapter. This energy is available to plants and animals through complex biological processes such as photosynthesis, respiration, and metabolism. These provide the raw materials for chemical reactions that help nearly all living things use energy resources to grow and reproduce. The way we humans live now consumes far more than the bare amount of energy required for survival. Will we learn to satisfy this need in a way that can continue for the generations to come? It is a difficult problem to answer, but that must not stop us from asking the question.

Practice Problems

1) A classic 60 watt light bulb (these are called *incandescent*) uses 60 joules of electrical energy every second. A *watt*, the rate of energy use, is equal to 1 joule per second. An LED (Light Emitting Diode) can produce the same amount of light using only 8 joules per second. There are 3,600 seconds in 1 hour.
   a) How many joules does the incandescent bulb use in 1 hour?
   b) How many joules does the LED bulb use in 1 hour?
   c) If you have 10 lights in your house, each on for 10 hours per day, how many joules can you save each day by changing them all from classic to LED lights?

2) Suppose that you use 75 liters of water to take a hot shower (that's about 20 gallons). Use the formula $\Delta H = m \, c \, \Delta T$ to calculate the joules of energy needed to raise the temperature of all of that water from 18 °C to 38 °C. That is a good estimate for the starting and final temperature of the water in your home. Remember, 1 Liter of water has a mass of 1 kg (1,000 grams), and the specific heat capacity of water is 4.2 J/g°C.

4
Heat and Light
4.1 Energy in Motion

We have explored the energy of moving molecules, which is the foundation of thermal energy. In a later unit, we will look at motion on a larger scale, such as a thrown ball or a moving car. In each case, these are forms of *kinetic energy* – the energy of moving mass. This chapter will look at the movement of *energy itself*. Of the many forms that energy can take, some can travel through other materials, or even through space, and transport from one place to another. Although it seems to be a subtle word-game to talk about energy *in* motion rather than energy *of* motion, there is actually a big difference.

If you have visited a large body of water, you have seen waves approaching shore. Small waves even occur in a bathtub or swimming pool. In a lake or ocean, the energy that causes these waves comes mainly from wind blowing on the surface; in the bathtub or pool, the energy comes from your movement in the water. In each case, the wave motion travels *across* the water, causing the water surface to move up and down. The water does not travel along with the wave; if it did, the middle of the ocean (or pool) would soon be empty!

This wave action is an example of moving energy, and a good model for many other types of waves. Energy can move from place to place by different kinds of wave, and we will often refer back to the example of water as we explore waves in more detail.

Consider a pool of water at equilibrium; that is, not visibly moving. A single jolt to that equilibrium can move through the pool as travelling energy called a **pulse**. More important to our current discussion is when the disturbance repeats in a regular pattern. This is called a **vibration**, which is the source of a wave. Stretch a rubber band between two fingers, and pluck it with the other hand. You can see it vibrate. Stretch it tighter, and you can hear a sound from the rubber band. You are not hearing the vibration itself, but tiny pulses of compressed air due to the rubber band moving quickly in a repeating pattern. This wave of compressed air is typical of all sounds.

The compressions move through the air in all directions, so anyone nearby feels these repeated pulses with instruments specially designed for the purpose...called ears. That is just what sound is: repeated pulses of compressed air that strike your ears, which your brain interprets as different pitches according to the rate of vibration.

Sound waves always travel through some substance. This material is usually air, but compressions can travel through many other materials. Put your ear next to a steel pipe, and ask someone to tap on the other end. The compressions travel even faster through steel than through the air. The material that a wave travels through is called the **medium**; but as we shall see, not all types of waves require one.

A single jolt of energy is called a PULSE. Repeated pulses are a VIBRATION, the source of all waves. Many, but not all waves, travel through a MEDIUM.

This famous painting by Japanese artist Katsushika Hokusai shows the frightening amount of energy unleashed by The Great Wave off the coast of Edo (now known as Tokyo). As we shall see, there are many kinds of wave.

4.2 Describing Waves

A smooth lake is not carrying wave energy across it; but drop a stone in the lake, and a ripple moves away in all directions. The water surface moves up and down as the energy passes, but it is important to remember that when we talk about a moving wave of any type, the energy is moving *across* the water, although the medium can move up and down as the energy passes through. A bigger stone gives the wave more energy, so the water moves higher up and down. The word for the amount of energy in a wave pulse is **amplitude**; for water waves, you can measure the energy by the height of the wave.

There are more interesting things to notice about this wave: first, the wave moves away from the dropped stone in a circle. That means it moves in all directions, as long as it has a path to move in all directions. It can't move up into the air where there is no water to disturb, and it can't move down toward the bottom because there is no surface where the water can deform easily.

Notice that the height of the wave gets smaller as the circle gets bigger. There are two reasons for that: a bigger circle means the energy is spread out more, so the *amplitude* decreases as the same amount of energy is distributed into a larger circle. Also, the water molecules moving against each other change some of that energy into thermal energy, so there is less energy to contribute to the wave itself (remember the conservation of energy law!).

If we keep dropping stones into the water, say, one per second, we will notice that the rings move away in a concentric pattern and they stay equally spaced as the rings get larger. This tells you that the speed of the waves does not change as they get farther away from the source and their amplitude gets smaller. The **wave speed** does not depend on the amplitude of the waves.

AMPLITUDE is the amount of energy in each pulse of a wave. The wave moves through the medium at a constant WAVE SPEED, which does not change for different amplitudes. A small wave moves just as fast as a large amplitude wave.

The spacing of waves also has a special name. For any type of wave, the distance from one pulse to another is called the **wavelength**. The rate that you drop the stones – in this case one per second – is called the **frequency**. The frequency of any wave is measured in pulses per second, called **hertz**. The interesting thing is that changing the frequency (say, dropping a stone every half-second) changes the space between pulses, but does not change the wave speed. Each pulse still moves across the water at the same rate.

The distance between any two successive pulse is called the WAVELENGTH, and the number of pulses every second is called the FREQUENCY.

Look carefully at the ripples in the water. The distance from the top of one wave to the top of the next is the *wavelength*. That length is the same for any two waves, and it stays the same as the waves spread out from the source in concentric circles, showing that the *wave speed* stays the same as well. If the speed decreased as the wave loses height or *amplitude*, the waves would bunch up as they get farther from the source.

Activity: Mini Hendrix

Materials: beach towel   rubber bands   paper or plastic cup

With a partner, hold the towel tightly at each end. One person should give the end a quick shake. A *pulse* of energy moves from one end to the other. Remember, the towel moves up and down as this energy passes from end to end. Repeat several times and observe.

1) What happens to the amplitude (height) of the pulse when you use more energy to shake the towel?
2) Does the speed of the pulse across the towel seem to change according to the size of the wave?

Stretch a rubber band around the paper cup from end to end Hold the cup in one hand; with a finger of the other hand, flick the rubber band where it crosses the top. Observe the vibration and listen for the sound; you may also feel the cup vibrating.

Flick the rubber band harder. That increases the *amplitude* (amount of energy) of the vibration. Then try squeezing the top of the cup slightly. Squeezing *across* the band stretches the band, and squeezing *in line* with the band loosens it. This changes the *frequency* of the vibration. Listen to how each of these actions changes the sound. Rewrite the sentences below, selecting the correct words from inside the parentheses in each case.

3) "Increasing the amplitude of vibration (increases/decreases) the (volume/pitch)."
4) "Loosening the stretch of the rubber band (increases/decreases) the (volume/pitch)."

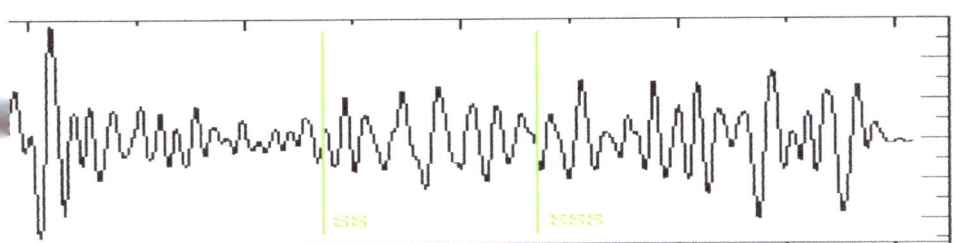

Part of the seismograph chart of the 1906 San Francisco earthquake. Seismographs measure vibrations of the ground. The earth is the medium that carries the vibration energy from the source (called the epicenter of the earthquake) to the city.

Practice Problems

1) Earthquakes are waves that travel through the earth, caused by natural geological processes. A seismograph can help scientists understand the destructive power of earthquakes. In the sample chart from a seismograph shown above, time is on the horizontal axis; amplitude is on the vertical axis.

   d) How many pulses of energy are recorded in the time between the green lines on the chart?

   e) The amount of time between the green lines is 0.02 seconds. What is the frequency of the earthquake vibration? Remember that frequency is number of pulses per second; divide!

   f) What does the chart show about the amplitude of the vibration? Is it the same for each pulse? How can you tell?

   g) Have you ever experienced an earthquake? If so, describe what it felt like. If not, describe what you *think* it would feel like.

2) Find a long continuous steel rail; a chain link fence or outside stair rail will work well. Have a partner stay at one end while you go to the other. Listen with your ear on the rail as they tap it with the stick, then listen for the sound through the air. Which is louder and more distinct? Can you tell which medium carries sound faster?

3) Amazingly, the human hear can detect sound frequencies as low as 20 pulses per second (20 hertz) up to about 20,000 Hz. Do an internet search for "frequency hearing test", and use it to test your own hearing. Can you hear that full range? It is likely that the speakers on your computer cannot reproduce all of these frequencies, so your test will probably be limited by that factor more than by your own hearing.

4) Mosquitos, bees, hummingbirds, and flies flap their wings at frequencies within the range of human hearing. To say that this makes a noise is simply to say that we can hear the frequency of air compressions this creates. Mosquito frequency ranges from about 400 Hz to about 800 Hz (females, which are the biters, are at the lower end of the range). Bees flap at a pretty consistent 180 Hz. Hummingbirds flap from about 40 Hz to about 70 Hz, depending on the species. Flies buzz at about 200 Hz, though this ranges up to 2,000 Hz, depending on size and activity.

   Have a partner search online for videos of each of these animals in flight. Without looking at the screen yourself, try to identify each one by the sound. You may also have an unpleasant feeling listening to a few of these, based on your past experience!

Extending Ideas

The speed of sound in air is about 330 meters per second. Use this information for problems 1 and 2.

1) If you are sitting in the bleachers at a baseball game, you can see the batter hit a ball before you hear it. If the sound takes ½ second to get to you, how far is your seat from home plate?

2) Sometimes people like to count the seconds after seeing lightning and hearing the thunder. If you hear the thunder 5 seconds after you see the lightning, how far has the sound traveled in that time?

3) Back in the day, kids would attach a baseball card to the frame of a bicycle so that the spokes flapped the card as they rode. The sound, of course, increases in frequency as the bike speed increases. Try it and hear for yourself!

The aftermath of the 1906 San Francisco earthquake.
The energy of an earthquake travels through the earth as a wave.

4.3 Charge Vibrations

The previous chapter described the three ways that heat is transferred. Convection occurs in fluids, when fast-moving molecules carry kinetic energy to other molecules. Conduction occurs when vibrating molecules jostle nearby molecules. Radiation also happens because of vibration, but on a much smaller scale. In this case it is not the molecules or even atoms that are vibrating, but the charged particles within the atoms - the electrons - that are responsible for the energy transfer. We need to dig deeper to understand how that works.

FORCE FIELDS can apply a force without touching. Although this idea seems magical, it is extremely common. The gravity field of earth pulls on a ball in your hand. A magnet can pull on a nearby paper clip.

First, let us recall work from even further back: unit 1 covered the concept of **force fields**; regions of space that can produce forces on particles nearby. Among the types of fields are *gravity fields*, which put force on any nearby mass (we call that force *weight*). *Electrical fields* put force on charged particles, such as the electrons and protons within atoms. *Magnetic fields* put force on other magnets or on materials that have a property called *ferromagnetism*. Under the right conditions, magnets can also put forces on charged particles.

What are the "right conditions"? In a word, motion. If a magnet is moving near a charged particle - or if a charged particle is moving near a magnet - they will put a force on each other. Sound strange? Well, it *is* strange. And it gets even stranger. The energy of magnetic and electric fields putting forces on each other can move through space, carrying that energy with them. This amazing phenomenon is called electromagnetic radiation, and it is responsible for heating by radiation, and for light, x-rays, sunburn, cell phones and many, many other phenomena. But as usual in science, you don't have to just *believe* anyone - even this book. You can test it for yourself!

Magnets readily stick to the side of a refrigerator, but not to the aluminum soda cans inside. Is it possible for magnets to put a force on aluminum?

Activity: Electromagnetic Slide

The first relationship we need to examine is the assertion that magnets put a force on charged particles, *but only if one or the other is moving.* Since everything is made of atoms, and atoms are made of charged particles, we can investigate this claim.

Materials: washer   ring magnet   aluminum sheet

One of the circles is a magnet, and the other is just a steel washer. The two look pretty similar, though.

1) Using just these two items, can you tell which is which? Spend a few minutes and try; describe your observations and results.

2) The aluminum sheet is non-magnetic. Using this, can you tell which is the washer in which is the magnet? Spend a few minutes and try; explain your results.

Prop one side of the aluminum sheet on a box or a pile of books. Place the washer and magnet at the top, and watch them slide down. If neither moves, make the aluminum ramp steeper.

3) Now can you tell which is the washer and which is the magnet? Describe the evidence and try to explain a reason this happens.

4) Do you think that a similar effect would occur if you use a piece of wood or plastic instead of aluminum? First predict (in writing), then try it and see! Describe your observation.

4.4 The Electro-Magnetic Link

The previous activity showed the relationship between electric forces and magnetic forces. As the magnet slides along the aluminum sheet, it puts a force on electrons in the aluminum. As a result, the electrons move. When electrons move, they make a magnet...which in turn puts a force on the original magnet! This force affects the permanent magnet– but only if it keeps moving!

People have noticed electric forces for thousands of years. The word *electric* comes from the Greek word for amber, *elektron*. Amber is actually fossilized tree resin, which can hold an electric charge when rubbed, like a balloon that is rubbed on your hair. The amber (or the balloon) can then put an electrostatic force on nearby objects. Likewise, magnetic forces have been known for thousands of years. The word *magnet* comes from a region of ancient Greece called Magnesia, a source of naturally occurring magnetic material.

Both electric and magnetic forces have been studied for many centuries, with some scientists suggesting there may be a link between the two. That link was not found until about two centuries ago. Since then, a number of scientists contributed to a growing pool of knowledge about the physical and mathematical relationships between electricity and magnetism.

This work led to the development of **electric generators, electric motors**, and eventually to the delivery of electric energy to most homes all over the world. It is easy to see that our modern lifestyle depends on this convenient source of energy. From a practical perspective, this is the biggest influence on our lives that came from understanding electromagnetic science.

A generator produces most of the electrical energy that you use every day. These devices usually have a coil of wire rotating within a magnetic field. This creates a force that pushes electric current through the coil. It is the same relationship between magnetism and electricity that you observed in the Electric Slide activity. Electric motors use the same relationship, but run it in reverse – so that electric current creates a force on a magnet that causes it to rotate. In fact, some electric motors can run backwards to work as generators.

> A GENERATOR changes energy of motion into electrical energy. A MOTOR changes electrical energy into energy of motion.

An electric *generator*, such as this one designed to provide emergency power to a home, uses a gasoline-powered engine that spins a magnet to make electrical energy. An electric *motor* is basically a generator in reverse, using electric energy to create a magnetic force to spin a rotor.

The energy needed to rotate the coil of wires can come from many sources. Emergency generators used in homes commonly use small gasoline-powered motors, similar to lawnmower engines. Giant electric generators used to supply a city might use coal, heat from nuclear fission, or natural gas to create superheated steam that spins a turbine. Renewable energy sources transfer the kinetic energy of wind or flowing water to spin a turbine.

Learning how to harness the relationship between electricity and magnetism brought profound changes to our lives. Manufacturing no longer depends on human, animal, or water power to run machines. Artificial light brightens our nights both inside and out. Many devices that were once powered by hand were converted to run on electric motors, and many new devices were invented. In the last half-century or so, information technology (IT) has grown from a few computers worldwide to powerful handheld devices that put the world at your fingertips, in the form of highly controlled electromagnetic devices. It is hard to imagine life today without electricity.

4.5 Sound and Light

The discovery of the interaction of electric and magnetic forces inspired incredible advances of the modern world. From a scientific perspective, perhaps a more important discovery happened in 1862 when James Clerk Maxwell published mathematical formulas that describe how the interaction of magnetic and electric fields work together to create and spread **electromagnetic radiation** (**EMR**).

EMR or ELECTROMAGNETIC RADIATION is produced by vibrating charged particles. EMR waves travel without a medium.

You have seen how different conditions can change the amplitude or frequency of sound waves. Consider this: when you listen to a band from some distance away, do you hear the high-pitched instruments first, the low pitched instruments, or does the sound from each get to you at the same time? All of the sounds travel at the same speed – if that were not the case, music would sound very strange.

What does that have to do with EMR? Electromagnetic waves are NOT compressions of air, like sound waves. EMR does not need air to move from one place to another, and in fact moves most freely through the vacuum of space. However, there are similarities.

EMR is caused by vibrations, which can occur at different frequencies. Like sound, EMR spreads in all directions from the source, carrying energy from the original vibration. Like sound, our bodies are equipped with sensors that can receive and interpret different frequencies of EMR. For sound, these different frequencies are called pitches, or musical notes. For EMR, the different frequencies represent different parts of the **electromagnetic spectrum**.

At the lowest frequencies, our skin feels the EMR as heat energy. At a small middle range of EMR, our eyes detect these frequencies as colors of light. Most of the electromagnetic spectrum is outside of the range that you can see, though we can sometimes observe its effects. One of the more unfortunate of those effects is all too familiar during the summer. Sunburn is the result of high-energy EMR damaging our skin!

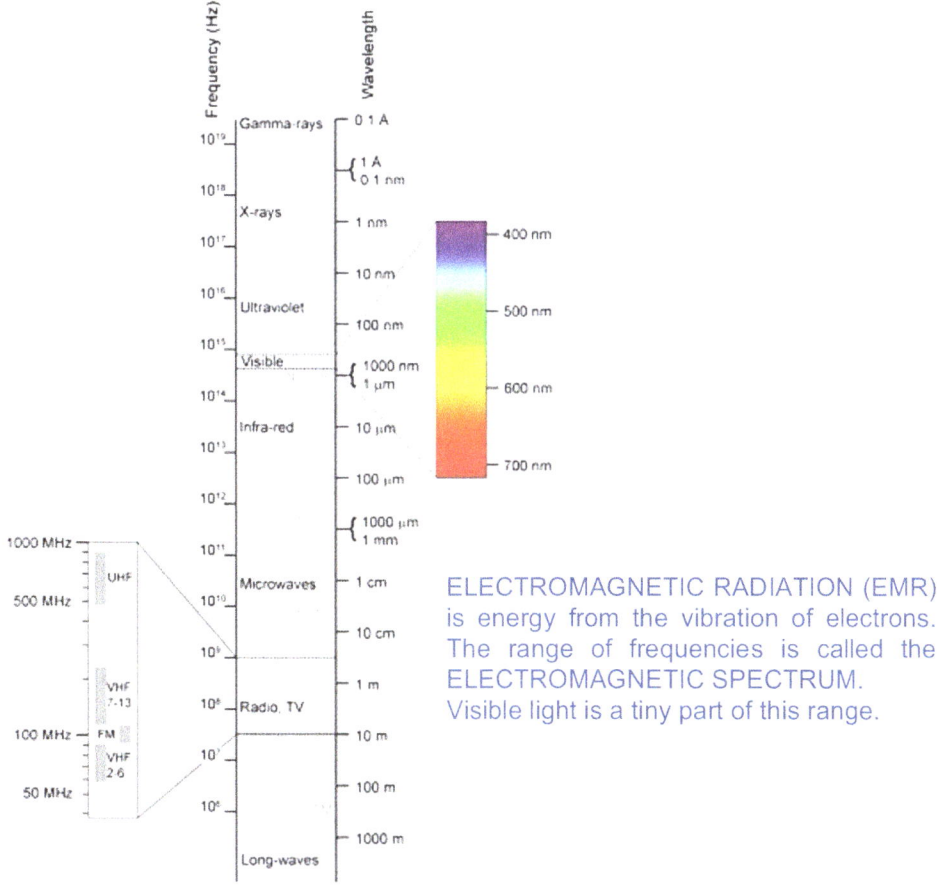

ELECTROMAGNETIC RADIATION (EMR) is energy from the vibration of electrons. The range of frequencies is called the ELECTROMAGNETIC SPECTRUM. Visible light is a tiny part of this range.

The source of EMR is not a vibrating object, but vibrating electrons. If you strum a guitar, all of the strings you touch will vibrate at their own characteristic frequency. In the same way, energy added to an atom will cause the electrons to vibrate. The frequency of this vibration depends on the electron arrangement, and that frequency is the frequency of EMR emitted. Since atoms of each element have a characteristic arrangement of electrons, each element emits a specific set of EM frequencies when energy is added.

The sun is made of mostly hydrogen and helium. Like all elements, they only emit EMR in a specific range of values. Life on earth has evolved to perceive these frequencies, which we simply call light, and to distinguish between them. Our eyes take in different frequencies, and our brains interpret these as colors!

Sound waves travel at the very fast speed of about 330 m/s in air – about 12 times faster than highway driving speed. But light and all other EMR travels almost a million times faster than that! The speed of light is 300,000,000 meters per second, sometimes called the speed limit of the universe. No object can travel as fast as light.

Radio waves are a type of electromagnetic wave. This EMR is amplified (increased) in the radio, then passes through a coil of wire inside a magnet. This forces the vibration of a thin paper cone in the radio speaker. The vibration creates a sound that travels through the air to your ears. Radio waves are *not* sound, but they use the electro-magnetic interaction to *create* sound!

Activity: The Light We Cannot See

Materials: microwave oven — a tortilla
microwave-safe dish — ruler
digital camera — remote control

You may have learned about the visible light spectrum in either art class or a previous science class. It is one of the many wonderful links between art and science! Visible light is the small range of EMR that our eyes can detect. The memory aid "Roy G. Biv" helps you recall the colors from lowest frequency to highest: *red, orange, yellow, green, blue, indigo, violet*. There is a wide range of EMR outside of that small set of frequencies, however. We will explore a couple of them here.

Part 1: Microwave ovens use EMR to cook food. The frequency is much lower than visible light, but of course you can feel the effects of the EMR energy whenever you heat food in a microwave oven. In this activity, we will measure the wavelength of microwaves and use that value to measure the speed of light.

If you place food inside a microwave, the highest energy EMR passes through it twice with each wave pulse (because it goes up and then down). The spots where the EMR passes through will absorb this energy as heat, but the rest of the food will stay relatively cool until heat travels by conduction or convection. Most microwaves use a turntable so different parts of the food are exposed to this high energy.

Remove the turntable (or flip it upside down), and place a tortilla (which does not conduct well) on the dish. Close the door and run the microwave on high for 30 seconds. Look for distinct scorch marks on the tortilla (without moving it); if none are visible, run the microwave for another 10 seconds. When the marks appear, remove the tortilla, and measure the center-to-center distance between two scorch marks. These are the high-energy spots of the microwave.

Because the microwaves go through the food twice with each pulse, the wavelength is actually two times the distance between burn marks. Multiply the distance by two to find the wavelength. Be sure to convert the units to meters.

Microwave ovens use a frequency of 2,450 MHz; that means the vibration is 2,450,000,000 (2.45×10^9) times per second. If you multiply the frequency times the wavelength, you will get the wave speed. The numbers are a bit unwieldy, but use a calculator and try it. The speed of light (and all EMR) is about 3×10^8 meters per second.

1) What did you get as the speed of light? Is it fairly close to 3×10^8?

Part 2: most remote controls use infrared light to send a signal. This light has a frequency just a little lower than visible light. Do not aim the remote directly at your eye, but look at the business end of the remote control at an angle as you click it a couple of times. You should not see any light. Now aim the camera at the remote. While looking at the camera view, click the remote. You may see the light – digital cameras are more sensitive to these frequencies than human eyes. Certain smart phones have an infrared filter on the front-facing camera that does not allow these frequencies to pass through; try the rear-facing camera instead.

4.6 The Speed of Dark

The moons of Jupiter were discovered by Galileo in 1609. In 1676, a Danish astronomer named Ole Roemer was tracking the motion of one of Jupiter's moons, Io (pronounced eye-oh). According to everything that was known about gravity, Roemer expected these orbits to always take the exact same amount of time. He was hoping that this regularity would help navigators and explorers use sightings of Io to help determine their exact location on the earth.

Roemer calculated that the timing of the Io's orbits changed depending on the positions of Earth and Jupiter in their own orbits around the sun. He realized that it was not the orbits that were irregular, but his observations. The problem was that it took time for light to travel across the great distances in space. Working from months of careful observations, he found the speed of light!

Galileo himself once tried to find the speed of light, using two lanterns, an assistant, and two hills a known distance apart. You can probably figure out his method; but Galileo was only able to measure his own reaction time! Roemer's method, corrected for distances in the solar system as we now know them, was very close to our currently known value for the speed of light: about 300,000,000 meters per second or 186,000 miles per second. Even if Galileo's hills were 10 miles apart (about the farthest you can expect to see a lantern), the time for light to go back and forth is about 1/100,000 of a second. His experiment would have had better results if he could have placed a lantern on Jupiter!

However you measure it, light travels awfully fast. Maxwell's equations showed that this is the exact same speed that all electromagnetic waves travel. That proved to Maxwell that light is just one part of the wide variety of waves that are all part of the electromagnetic spectrum. The word *spectrum* is also used to describe the range of colors of visible light; but as we have observed, the full range of EMR is much greater than the small bit that we see!

> James "Cool Papa" Bell was one of the fastest runners to ever play baseball. His teammate Leroy "Satchel" Paige once noticed that the light switch in their hotel room didn't work right, so that there was a delay between flicking the switch off and the light turning out. Satchel, a great story-teller, used this experience to tell people that Cool Papa was so fast that he could turn out the light and be in bed before the room got dark!

4.7 Doing the Wave

Now that you have observed several different kinds of wave, let's see how they can behave in ways that might surprise you. Some of these "tricks" may be familiar in that you have observed them, but you may not realize that they are caused by the unique way that waves interact with the world.

REFLECTION is when a wave reverses direction at the boundary of a medium.

For example, all types of wave can **reflect**. You have certainly seen the reflection of light waves in a mirror. You have probably heard the reflection of sound waves off a distant hard surface, which we call an echo. Water waves can reflect from the side of a bowl or bathtub; on a larger scale, waves at the beach will reflect from the side of a pier. You need to observe closely to notice some of these reflections.

Look even closer, and you will notice more. Do sound waves reflect from nearby surfaces as well as distant ones? The answer is yes, but it is usually hard to notice in a room, because the echo is **absorbed** by soft materials such as carpets, drapes, furniture, and even people. This is why most concert halls and theaters have fabric on the walls and cushioned seats: you would not want echoes interfering with the sound. If you have ever been in a cave, or a room with only hard surfaces and very few people, it is difficult to have a conversation. The voices are muffled and murky because of all the reflected sound waves.

ABSORPTION means that wave energy dissipates into heat energy.

Reflection of light from a smooth water surface can produce a spectacular image. Go on a photo hunt and try to capture a reflection from water!

The fact that waves can be absorbed or reflected produces more familiar effects. When light hits a red rose, most of the visible frequencies are absorbed. This absorbed energy warms up the rose. But some frequencies are reflected: those in the small range that we perceive as "red." So to say an object is red…or blue, yellow, or green…is only to say that the object reflects that small range of visible light that we perceive in that color. How about a white object? It reflects many frequencies, which our eyes blend together. Color sensing cells in our eyes – called cones – are all stimulated, and our brains perceive this blending of many frequencies as white. At the other extreme are objects that absorb light of many frequencies. Since no visible light is reflected, we see black.

The wooden bridge is painted in colors that reflect light waves that combine to make purple.

The aquatic plants reflect frequencies that our eyes perceive as green.

Finnegan's fur absorbs all frequencies of visible light. Also, he's a very good boy.

Absorption of any kind of wave causes objects to heat up. Again, this is conservation of energy; the wave energy does not disappear, but is converted. Have you ever walked on a dark surface, like a paved road, in bare feet? As the sun shines on it, that road gets hot! If you walk on a painted white stripe, you will find it is noticeably cooler, because more EMR is reflected. Another good reason to make sure that you cross on the crosswalk!

Energy travelling as a wave can transmit to an object. Place a coin in the middle of a desk. Now slap the edge of the desk. Can you make the coin jump? The energy of your slap makes the desk vibrate, and that energy travels as a wave across the surface to the coin. The vibration of the desk creates a sound, as well. The noise of your slap travels through the air to your ear. Both of these are examples of energy travelling as a wave, then interacting to give energy to the coin, or to signal a sound wave to your brain.

Waves can carry energy (transmit) from one substance to another. If the wave travels at different speeds in those two substances, the direction changes where one substance ends and the other begins. This is why a straight stick half submerged in water appears to bend suddenly at the surface. This effect, called **refraction**, is how eyeglasses, telescopes, and other lenses redirect light in various ways.

REFRACTION is a change in the direction of a wave as it transmits from one medium to another. Refraction can also be a more gradual change in wave direction within the same medium.

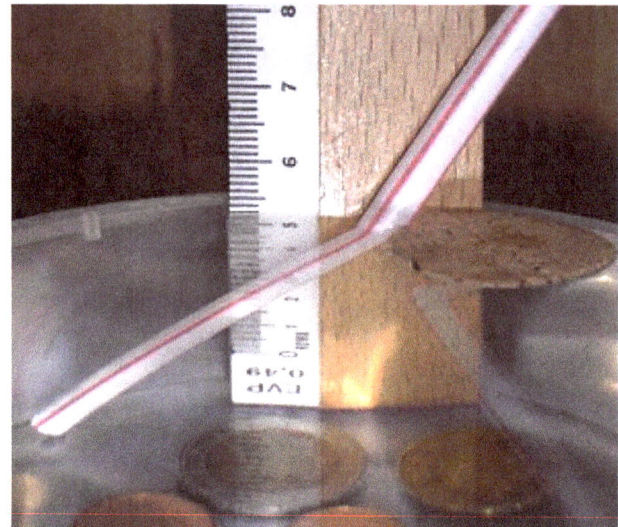

The straw seems to bend as it emerges from the water. The submerged part of the scale appears to be compressed. Both of these effects are caused by *refraction* - light bending at the water surface.

The direction can also change gradually; this is also refraction, and it happens when the wave travels through a substance that changes in ways that slow down the wave speed. This happens to water waves as they approach the shore. The shallow water causes the wave to slow down and curve toward the beach. Changes in air temperature cause changes in the speed of sound that you may notice as sound travels over a large body of water on a summer evening.

Some light frequencies refract more than others. Sunlight, composed of many frequencies, enters a raindrop, and red frequencies bend less than blue frequencies. Light travels through the raindrop, refracting at both curved surfaces. The different amounts of refraction cause the light frequencies to separate (called *dispersion*), red from blue and all of the colors in between. We are rewarded with a rainbow!

A double rainbow is one of the most spectacular examples in nature of the amazing behavior of waves. Be glad that you live in a universe where rainbows happen, and that your eyes evolved with the ability to see them!

Sunlight enters our atmosphere after travelling through space. Some of the high-frequency visible light is scattered by molecules of gas in the atmosphere. We see these high frequencies against the lightless backdrop of space when we look up. That is a scientific answer to the question often asked by children, "why is the sky blue?"

But the sky isn't *always* blue. In the morning and evening, the sun is low in the sky. Sunlight must travel through a thicker layer of atmosphere to reach your eyes. That means more frequencies of visible light are scattered, and the sky may appear orange, yellow, and red. This effect is increased by particles in the air, such as clouds. It is always worthwhile to pause and gaze at a spectacular sunset!

If you look closely at a soap bubble, you may see a range of colors; the same effect is visible in a thin film of oil in a puddle in a street or driveway. These colors result from light frequencies cancelling one another out – a wave behavior that is also partially responsible for the second rainbow in the double rainbow on page 119.

Waves can create even more unexpected and interesting effects. When you know what to look for (and listen for), you might observe these effects any place there is sound or light – and that's almost anywhere you are likely to go!

Paris is known as "the City of Light," a nickname that predates the invention of electric lighting. This photo taken from the International Space Station shows this is an apt description. Electric lights have changed the look of the world at night. Whether this change is good or bad is a matter of opinion.

Extending Ideas

1) All sounds are started by vibrations. Research exactly what is vibrating to produce a sound from each of these examples.

    a) a guitar
    b) a piano
    c) a saxophone
    d) a drum
    e) a speaker
    f) a singer's voice

2) Electrical energy in the United States comes from many different sources. Research the sources most used in your state, and create a pie chart showing the prevalence of these different sources.

3) Section 4.3 shows a range of electromagnetic radiation frequencies including gamma rays, x-rays, ultraviolet, infrared, microwaves, radio and TV waves. Select one of these and research a typical source and common use.

4) When you look at a rainbow, you see colors as a result of refraction and dispersion. When you look at a *picture* of a rainbow, you see colors because of reflection and absorption. Write a paragraph explaining this apparent contraction.

5) Some animals can see light in the infrared ("below red") range, and other animals can see light at the other end of the visible spectrum, ultraviolet ("above violet"). Research to find out which animals, and what features of the world they see that we can't.

4.8 Analog vs. Digital

We have already seen that sound travels through the air (and through other materials) as a compression wave. When these air compressions encounter your ear, the thin membrane called your eardrum vibrates in response, sending a signal of loudness and pitch (frequency) to your brain. If these sound waves strike a microphone, they cause a vibration that converts the mechanical wave to an electromagnetic wave, which travels at the speed of light. Visible light itself travels as a wave, which our eyes receive and decode as visual signals. Sound and light – our two most trusted links to the outside world – convey information through waves.

But there's more. Television and radio broadcasts use EMR. Wi-Fi is a short-range electromagnetic wave adapted for computers. The signals to and from our cell phones (called "4G " or "5G" because they are the fourth or fifth generation of the design) use EMR as well; similar to Wi-Fi, but designed for long-range signals.

Most information comes to us through waves that are produced by some physical action of vibration. This is called an **analog** signal, because it is *analogous* to the vibration itself. Vibrations are continuous movement, and the signals (waves) produced by these vibrations are continuous as well.

ANALOG: a continuous signal, like the source that creates it.

DIGITAL: a binary (on or off) signal, stored and processed in electronic devices.

A problem occurs when we try to record this information or process it using a computer. Recording and processing devices use electrical signals. These are either "on" or "off", with no in-between – so they are not continuous signals. Think of a swinging pendulum, which is a sort of slow vibration. The pendulum travels from one side to the other in a continuous motion. An electrical signal cannot see this motion – it only sees the pendulum on one side or the other. This type of information is called a **digital** signal.

Digital signals are used in all kinds of electronic devices. You may have heard the term *binary* applied in this sense – it means there are only two possibilities: on or off. These are represented mathematically as 1 or 0, or visually as dark or light pixels. In fact, your television or computer screen, which may look like a continuous signal, is actually a

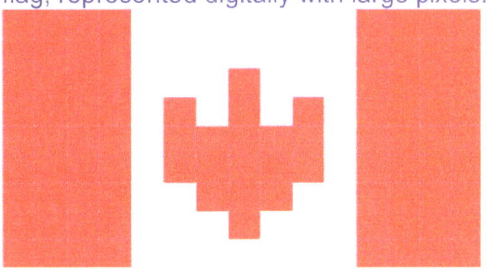
The Canadian national flag, and the same flag, represented digitally with large pixels.

set of very small pixels that appear continuous at a distance. This is obvious in old-style video games. The limitations of computer power required programmers to use fewer pixels, larger in size. These were obvious even from several feet away. Modern high-resolution video games use many more pixels, which are harder to see, even from a short distance. The pixels are still there, however, and the requirements of digital devices mean that every pixel is either on or off.

The most important advantage of digital signals is the speed of processing the information using electronic devices. Storing information for later retrieval is much more reliable, as well. All of your computer memory, movies (either streaming or on DVD), and music is stored, retrieved, and processed from a digital signal. As shown in the flag example, some details are lost in translation, but we make up for that with certainty. In other words, there is no ambiguity or questions about the signal; it is either on or off, not in-between. Improving technology has continuously increased the precision of reproduction from the original (analog) signal.

Records (sometimes called *vinyl*) store and play back music using the vibration of a needle in a specially-cut groove in the plastic material. Some music fans are certain that this analog method reproduces music more faithfully (with higher *fidelity*) than digital processes.

Art / Photograph Credits

3 (crane): Art Explosion

5 (couple eating lunch) Wikimedia Commons; Bill Branson

8 (boat winch) Art Explosion

10 (hot plate) By GOKLuLe 盧樂 - Own work, CC BY-SA 3.0,
https://commons.wikimedia.org/w/index.php?curid=17050615

10 (gas burner) Creative Commons; https://www.flickr.com/photos/sfllaw/8753914/

16 (hot air balloon photo) Wikimedia Commons; Jakub Hałun - Own work, CC BY-SA 4.0, https://commons.wikimedia.org/w/index.php?curid=91131847

18 (Joule's apparatus) Wikimedia Commons; By Unknown author - Harper's New Monthly Magazine, No. 231, August, 1869., Public Domain, https://commons.wikimedia.org/w/index.php?curid=1527228

21 (colored samples), 22 (popcorn) Peter Swan

26 (thermometer scales) TheVovaNik, CC BY-SA 4.0 <https://creativecommons.org/licenses/by-sa/4.0>, via Wikimedia Commons

28 (NASA test firing) courtesy NASA

30 (molecular motion) Wikimedia Commons; Donado F, Moctezuma R, López-Flores L, Medina-Noyola M, Arauz-Lara J

31 (hot tea) Wikimedia Commons; By Ssgapu22 - Own work, CC BY-SA 4.0, https://commons.wikimedia.org/w/index.php?curid=89261341

33 (thermostat) Wikimedia Commons; By Flickr user midnightcomm - https://www.flickr.com/photos/midnightcomm/447335691/, CC BY 2.0, https://commons.wikimedia.org/w/index.php?curid=5596839

39 (faucet) Miguel Andrade, Public domain, via Wikimedia Commons

51 (unfrozen pond) Wikimedia Commons; By Dot Potter, CC BY-SA 2.0, https://commons.wikimedia.org/w/index.php?curid=9207758

52 (bottle and thermometer) Peter Swan

56 (volcano); 60 (beach scene) Microsoft

61 (condensation) Peter Swan

65 (metal heating) By PowderBN - Own work, CC BY 4.0, https://commons.wikimedia.org/w/index.php?curid=79371732

67 (Expansion joint) Peter Swan

68 (hot air balloon video) Wikimedia Commons; Samu / CC BY (https://creativecommons.org/licenses/by/2.0); link: https://commons.wikimedia.org/wiki/File:Hot_air_balloon_in_Spain.webm

71 (convection from hand) By Gary Settles - Own work, CC BY-SA 3.0, https://commons.wikimedia.org/w/index.php?curid=29523610

72 (double-walled vessel) By Gustav Robert Paalen - https://www.google.com/patents/US950557, Public Domain, https://commons.wikimedia.org/w/index.php?curid=54637939

73 (frying pan) By Evan-Amos - Own work, Public Domain,
https://commons.wikimedia.org/w/index.php?curid=11960778
74 (bulb) By LPS.1 - Own work, CC0,
https://commons.wikimedia.org/w/index.php?curid=29802905
77 (flashlights) Microsoft
78 (shoreline) Microsoft
83 (baseboard radiator) Public Domain,
https://commons.wikimedia.org/w/index.php?curid=1189349
83 (standing radiator) By Schorle - Own work, CC BY-SA 4.0,
https://commons.wikimedia.org/w/index.php?curid=75294903
84 (air-cooled huts) Microsoft
86 (drink) Microsoft
87 (steam train) Art Explosion
95 (Great wave) After Katsushika Hokusai - Restored version of File:Great Wave off Kanagawa.jpg Public Domain,
https://commons.wikimedia.org/w/index.php?curid=5576388
97 (ripples on a lake) By Garrett Sears garrettsears -
https://unsplash.com/photos/rXVFCA3fQ4IImageGallery, CC0,
https://commons.wikimedia.org/w/index.php?curid=61896156
99 (seismograph) By Pekachu (talk) - USGS pages [1], USGS page, Public Domain,
https://commons.wikimedia.org/w/index.php?curid=34091025
101 (San Francisco after earthquake) By Chadwick, H. D - cataloged under the National Archives Identifier (NAID) 524396., Public Domain,
https://commons.wikimedia.org/w/index.php?curid=2128556
103 Refrigerator magnets: Insa 4gp groupe ferro, CC BY-SA 4.0
<https://creativecommons.org/licenses/by-sa/4.0>, via Wikimedia Commons
106 (generator) Peter Swan
 https://commons.wikimedia.org/w/index.php?curid=40742899
109 (EM spectrum) https://commons.wikimedia.org/w/index.php?curid=4051422
111 (visible spectrum) By Maulucioni, - Own work, CC BY-SA 4.0,
115 (Taj Mahal reflection) Antrix3, CC BY-SA 4.0
116 (Finnegan) Dianne Crocker, used by permission
118 (refracted straw) Rainald62, CC BY-SA 3.0
<https://creativecommons.org/licenses/by-sa/3.0>, via Wikimedia Commons
119 (Rainbow) IIP Photo Archive, CC BY 2.0
<https://creativecommons.org/licenses/by/2.0>, via Wikimedia Commons
120 (Paris, night satellite image) NASA, Public domain, via Wikimedia Commons
123 (Canadian flag) E Pluribus Anthony / User:Mzajac, Public domain, via Wikimedia Commons
123 (pixelated flag) Sanchit Gupta, CC0, via Wikimedia Commons
124 (Record player) Franz van Duns, CC BY-SA 4.0
<https://creativecommons.org/licenses/by-sa/4.0>, via Wikimedia Commons

Index

		defined	illustration
absorption		115	
analog signals	122-124	122	124
Brownian motion	20		
calorie	22		
conservation of energy (Law of)		6	
convection	71-72, 76-78	71	71, 72
cooling systems		84	84
digital signals	122-124	122	123
Einstein, Albert	20		
energy	*throughout*	2	3, 8
chemical		5	
conserving	90-91		
electrical	105-107	5	
kinetic		27	30
mechanical		5	
nuclear		5	
of living things	22, 91		
potential		2,3	
radiant		5	
sources	81	82	
thermal	31-34	5,6	
transformations			8
entropy		40	
equilibrium	40-41	32	31
electromagnetic			
radiation	109-114	108	
spectrum	109, 111, 114	109	109
exothermic		22	
Fahrenheit, Gabriel	26		
food energy	22-24		

			defined	*illustration*
force fields			102	
generator		105-107	106	106
heat		1	6, 27	10, 16
	caloric theory	16	17	
	capacity of water	78-79		78
	engines	87-88		87
	exchangers		82	83
	specific	44, 47 table	34	
	units of	16-17, 22	18	
heat transfer		71-79		
	conduction	76	73	65, 73
	convection	71-72, 76-78	71	71, 72
	on earth	76-79		78
	radiation		74	74, 77
insulators			73, 84	
Joule, James		17-18, 26		18
medium			94	
metabolism			22	3
motor		105-107	106	
phase changes		50-58, 85		51, 52, 56, 86
	condensation		55	61
	crystallization	54	55	51
	evaporation		85	
	fusion	54	55	
	heat of table	58		
	sublimation		56	
	vaporization		55	56
pulse			94	
reflection			116	115, 116
refraction		117-119	117	118

		defined	*illustration*
temperature	1	27	
changes	31-39	35	31
scales	26-29		26
thermal expansion	65-68	66	67, 68
of water	79		
thermodynamics	60-	61	
Laws of	75		
thermometer	26		26
vibration	93-97	94	
of charge	102-103		
wave	93-97		95, 97
amplitude		96	
frequency		97	
speed		96	
wavelength		97	
work	2		2
working fluid		82, 85	

www.ingramcontent.com/pod-product-compliance
Lightning Source LLC
Chambersburg PA
CBHW060846220526
45466CB00003B/1261